UNIMAGINABLE HARVEST

HOW DID IT HAPPEN?

DR. FRANCIS N. KATEH

WORDS MATTER
P U B L I S H I N G
OUR WORDS CHANGE THE WORLD

Words Matter Publishing
P.O. Box 1190
Decatur, IL 62525
www.wordsmatterpublishing.com

ISBN 13: 978-1-962467-39-1

Library of Congress Catalog Card Number: 2024946124

Table of Contents

————·◆◆◆◆◆·————

Dedication

———◆◆◆◆◆·———

This autobiography is dedicated to my late Mother, Edith Hneayeneh Wallace. Despite her affliction, she never relented in making sure that her son didn't turn her infliction into a lackadaisical attitude toward attaining the desire to be a medical doctor. This story of her life has been manifested into something positive that may serve as an encouragement to those whose parents or friends have been stigmatized because of an infliction.

Hence, my first love, guarding angle, and interceder remembered our many discussions when you were here with me. Please be assured that as long as God continues to protect and give me a long life, I will do my part to ensure those discussions come to fruition.

Therefore, this book is dedicated to you and others who have gone through what you did; hence, ten percent of the proceeds of this book will go towards the health center named in your honor and scholarships for deserving students.

Acknowledgment

———·◆◆◆◆·———

Appreciating all those who inspired you could be challenging. Still, I hope those who may feel disenchanted for their names not being mentioned should please accept my apology. With that, let me begin by acknowledging the late Dr. Jim Goulding, who first encouraged me to write my story, followed by the late John Doyle, who, despite his financial support to me, continued to enable me to write the story not as an appreciation him, but allowing others to read and be encouraged that God is still on the throne and is able the make the impossible, possible. The late Bettie Wilson Story, editor of "The Reporter" of the Central Illinois Conference, now Illinois Great Rivers Conference, heard my story regarding my inability to get admitted at two medical schools because I wasn't eligible to earn any federal loan or support to enter medical school because of my F-1 Status. She wrote an article on me that led to 23 Churches committing to support me through medical school. A Special thanks to the Board of Trustees of MacMurray College and Central Illinois Conference, now Illinois Great River Conference, who supported me with their resources, building the foundation for what I am today. To the late Dr. Paul Farmer, founder of PIH (Partners in Health), the first time you met me and wanted to know my reason for becoming a physician, you indicated that I needed to write a book and you would have it published; unfortunately, you are not around, but your words keep echoing in my mind, and today, I want to say thanks, the manuscript is complete and published.

Writing the manuscript is one thing, but finding the appropriate name that epitomizes the message becomes critical; hence, I which to express my thanks and appreciation to Evangelist Daniel Arthur Moe, commonly called "Danny," who, through Devine intervention, spoke on the theme "Unimaginable Harvest" captivating my thought and immediately after his exhortation in Vancouver, Canada, I approached him. I asked if I could use his theme as the title of my manuscript, and he agreed. Today, my life journey is entitled "Unimaginable Harvest". Marj Ratel, Dan's Sister and founder of the Korle Bu Neuroscience Foundation, I am grateful for your role.

A Special thanks and appreciation to Rev. Dr. Levi C. Williams for proofreading the manuscript and the advice given to me that shifted the content of the book. "Big Brother," I will always be grateful to you.

Finally, to my dear wife, Katherine Marks-Kateh, and all of our kids, thanks for your encouragement and the pressure placed on me to have the manuscript complete and ready for publishing.

Tammy Corwin and your team at Words Matter Publishing Inc., thanks for the professionalism and creativity you have brought into bringing what was thought to be a herculean task to fruition.

Foreword

———·◆◆ ◆ ◆◆·———

by Rev. Dr. Arthur Flomo Kulah, Bishop Emeritus,
United Methodist Church, Liberia Annual Conference

I am filled with joy as I retrospect on a young man that I met in a village a year prior to being elected as Resident Bishop of the United Methodist Church, Liberia Annual Conference.

As an educator at the Gbarnga School of Theology, Gbarnga, Bong County in central Liberia trying to prepare the future messengers of the Gospel of Christ, I felt that there was a necessity to reach the rural towns and villages of Liberia in spreading the Gospel, and this had to be done through preparing the lay speakers in understanding the liturgy of Methodism. The growth of the Church was increasing exponentially, and the traditional process of ministerial education wasn't meeting the demands of evangelistic growth. As a result, I would go to various districts of the United Methodist Church to conduct training. One of such trips for training took me to Cape Palmas District and to be specific, Karlokeh, Maryland County. I was being hosted by my friend and Brother in Christ, Moffat Kateh-Mah Brown, and his wife. It was during that visit that I came across this lad who was so humble and voluntarily decided to do my laundry and ran errands for me. Not only that, but he would also come and quietly sit at the back of the room, listening to my training sessions. This caught my attention, and for the few days I stayed at his parents' residence, I saw his closeness to his mother. The short time I stayed left an indelible mark on me that made me feel that this young man will become something in the future.

I do remember about two years after being elected as Bishop, I received a letter from someone who didn't just want a handout but was willing to volunteer in order to find his niche in the healing ministry of the Church. I was amazed and quickly realized that assisting this young man in achieving his dream could be a benefit for many in the future. I honored his request to go to Ganta United Methodist Hospital as a volunteer. Over the years, I followed this young man as he pursued his dreams because of his relationship with his mother and the unfortunate situation of his mother's losing one of her eyes.

As a Bishop who was in charge of leading the shepherds of the United Methodist Church, it became obvious that investing in this young man will yield a positive dividend not only for the Church but for the people of Liberia and the rest of the world because of his genuine love for his mother, the church, and people in general. The bond that we share is not just as a Bishop and a church member but as father and son. This bond has manifested itself into so many positive attributes that have had a transformational impact in the healing ministry of the church and the country as a whole.

I am pleased that in spite of the many young people that I had to shepherd, my "son," Dr. Francis N. Kateh, has continued to listen to the directives of God in serving humanity with honesty, sincerity, but most importantly, obedience in the reliance of what God wants him to do. I have observed that he is not moved by worldly possessions which was made evident when he returned to Liberia immediately after his graduation from medical school abroad. Despite the instability in Liberia after the civil crisis, he returned to provide medical attention to those that were sick at a time when many others refused to return after being sponsored by the Church. Dr. Kateh made a difference and upheld the promise he made to God and the Church, and today, his involvement in the healing ministry has saved countless lives.

I am proud that my involvement in the preparation for what could come has set a path for greater things. "Unimaginable Harvest" epitomizes how this young man from a humble village in Liberia has become a valuable resource contributing not only to the Church but to humanity in general.

CHAPTER 1

———— ·◆• ◆ •◆· ————

Unexpected Arrival

Every morning, Edith Wallace woke up to start her daily routine on the farm. She tilled the soil to plant rice, corn, and cassava, which would eventually yield an abundant harvest to feed her large family. It was quite fascinating how Edith's husband, Moffat Kateh-Mah Brown, managed to house so many extended family members. At one point, twenty-three people lived under their roof. Their house, built in 1975, had six bedrooms, a dining room, and a large living room. The original house, located in a part of town called "Down Town," included well-known families like the Timothys, Brewers, Sackeys, and Siehs. The first house Edith and her husband built was made of mud and dried leaves. It had three bedrooms and a living room that also served as a courtroom, as Mr. Moffat Kateh-Mah Brown was not only a teacher but served as Justice of the Peace and Stipendiary Magistrate.

On the morning of July 12th, Edith's routine took an unexpected turn. After preparing the children for school, she began her journey to the farm. As she got closer, she felt a sudden rush of fluid down her leg—the rupture of her amniotic sac. One of the older boys living with the family was traveling with her. In Liberia, households often included extended family members beyond just siblings and offspring. Edith quickly instructed him to hurry back to town and call the Licensed Practical Nurse (LPN), commonly referred to as the "Doctor." As the young man jogged towards town, Edith slowed her walk, hoping to minimize the distance. However, her journey back to town was cut short as her cervix became fully dilated, and the baby's head emerged.

Edith left the main path and headed towards a large sassywood tree not far from the main highway to Barrobo. Barrobo was an indigenous town inhabited by the Bush-Grebo tribe, part of the larger Grebo tribe of Maryland County. The term "Bush-Grebo" distinguished them from the "Seaside Grebos" along the coast. Although the Bush-Grebos had many towns, one of the most prominent towns housed American missionaries from the Assemblies of God Church, who had built a refuge center for lepers near a village called Nowakeh.

As Edith squatted to get comfortable, the baby arrived, and a downpour of rain immediately followed. After a short while, she was relieved to hear the voice of the LPN. She called out, "I am right here." The nurse swiftly cut the umbilical cord, delivered the placenta, and wrapped the baby, ensuring the contraction of the uterus with gauze packing. They then gradually walked back towards town.

News of Edith's delivery spread quickly. Soon, she and the nurse heard jubilant songs as women from the community came to aid them, singing and dancing for the birth of Edith's bouncing baby boy under the culturally significant sassywood tree. This tree's bark was traditionally used in trials against those accused of crimes, especially witchcraft. The birth of the boy under the sassywood tree became as significant as the tree itself. In gratitude, Moffat Kateh-Mah Brown named his first boy child after the LPN, Francis Theophilus Nah Nyemoh; thus, the baby boy became Francis Theophilus Nah Brown. This is how Edith's son came into the world, according to her and the LPN.

The village of Karlokeh, now a major town, was ideally located and known as the "Gateway" to Maryland County. Surrounded by luxuriant tropical rainforest, it earned the nickname "California," representing the

State of California, USA, in Liberia. The town was founded by Chief Musu, named after a tree (Karlo- and -keh, meaning "rest" or "reside") in the local vernacular located in the middle of the town.

**Chief Musu's House with his grave and that of his Clerk,
Oldman Doyen where the Tree was located**

Karlokeh's strategic location made it a business hub. The road from Monrovia, Liberia's capital, to Haper City, Maryland County passed through the town, making it a bustling marketplace on Saturdays. People from nearby towns and cities like Barrobo, Nyenebo, Webbo, Pleebo, and Harper traveled to sell and purchase produce. By early afternoon, sellers and buyers loaded their purchases into various vehicles to head back to Pleebo or Harper. Karlokeh, now elevated to a city, served as the headquarters of the Karluway District.

The entry to the beautiful (Town) now city of Karlokeh from Monrovia

CHAPTER 2

Edith Hneyanneh Wallace

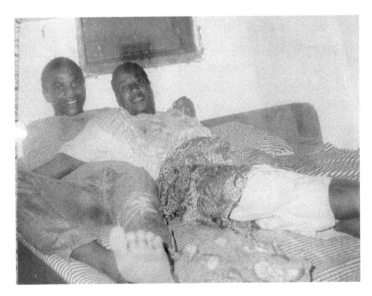

The only photo that I know off, of the late Edith H. Wallace and I taking on the morning of Xmas 1988, the last time I saw my beloved friend and mother

Edith H. Wallace had a remarkable life journey. Born to Kaleblasie Wallace and Mary Mahnoh Freeman, she was practically raised by her extended family after her mother left her as an infant, relocating to the Ivory Coast in search of greener pastures. Her father's life was tragically cut short, leaving her practically an orphan.

At a tender age, Edith contracted bacterial conjunctivitis. With no doctor available, an herbalist placed herbs in her right eye, and she instantaneously lost it. This subsequently prohibited her from achieving an education due to the stigma associated with the purulent fluid running from her eye. She married Mr. Collins, but when their union did not produce a child, he decided to divorce her. What some see as disgrace can sometimes be the beginning of grace. The local chief handling the divorce case fell in love with Edith, and she became one of his wives. Fortunately, she conceived within several months and gave birth to a girl. Her joy was cut short a day after her first daughter's first birthday. According to the story, a big birthday party was held, and it seems the toddler was poisoned, resulting in her death.

After a period of grief, Edith was blessed with another pregnancy and gave birth to another daughter. As the youngest among the wives, envy grew. With her infant approaching her first birthday, Edith learned that her mother had returned to Liberia and was residing in Karlokeh. Devastated by the envy and the incident with her first daughter, she decided to find her mother, whom she didn't know but who had returned from the Ivory Coast after her search for greener pastures proved elusive. Edith Hneayeneh Wallace was her only child.

As I sit reflecting in my home near Lake Shepherd, one of two lakes in Liberia, my thoughts drift to the stories I was told and those I remember. My late grandmother moved to Karlokeh because it was becoming a booming town with the potential for prosperity. People from different parts of the country were relocating there, turning it into a hub for entrepreneurs. This compelled the government to establish a school for the younger ones in anticipation of population growth. Moffat Kateh-Mah Brown was tasked with establishing this school. His journey, both uneventful and eventful, shaped my own story.

Edith arrived in Karlokeh to reconnect with her mother and find closure for the various calamities and personal, familial vacuum in her life. Several weeks after her arrival, she met Moffat, the man tasked with the herculean job of establishing a school that would serve as the foundation for the country's future. As President Tubman began a massive development agenda to integrate indigenous Liberians and Americo-Liberians, education became the cornerstone of this integration. A nation that invests in education is investing in its future, wealth, and sustainability. Moffat met Edith, an illiterate woman, and a relationship began that would lead to the reason for this book.

Edith was a committed, patient, dedicated humanitarian, and most importantly, she was God-fearing. She was so committed to her relationship with my father that I became her defender. I recall the last and only time I challenged my father, in 1988, when I went home for vacation from Monrovia. After driving for nearly twenty-four hours, I arrived around 1:00 AM. My mom and the rest of my siblings and extended relatives welcomed me. I didn't see my dad, so I assumed he was out of town. After the welcome, we all went to bed, and I slept in my mom's room.

By 6:00 AM, she woke up to light the fire in the kitchen outside the house and prepare hot water for tea and a bath. When she returned, I asked, "Mom, why were you out?"

"Oh, I went to prepare water for your dad."
I was curious. "Is he here?"
"Yes, he is, and in his room."
"Ok. So, he didn't come to see me when I arrived?" I asked.
"Oh, my son, your dad will not change."

Puzzled, I thought about what she meant. My dad went to take his bath, and I saw my mom going out to prepare more hot water.

"Who are you going to prepare water for this time?" I asked.
"Your dad's friend," she replied.

It was around 7:00 AM. I immediately went into my dad's room, took his mistress's bag, threw it outside, and told her to leave the house immediately. Dad, hearing the commotion, came in angry and wanted to hit me. I told him that if he hit me, I would defend myself as a man. He immediately began to blame my mom for my actions. I told him it was absolutely wrong to blame her.

"As a child, I witnessed your continuous disrespect towards Mom and couldn't say a word. Now that you're both older, you continue the same behavior. I will not accept this anymore. I have seen Mom in tears and would sit with her, hoping she wouldn't cry. When she went into that sad mood, I would cry with her without knowing why she was grieving. Since she won't express how hurt she is, I will defend her."

This was the level of her commitment to my dad. As a compassionate and kindhearted person, my mom was generous to the extent that the entire family couldn't understand why she would often give out the little we had to others. For example, when most of the traders in Karlokeh, especially those from nearby Barrobo, were stranded after business hours and all commercial vehicles had left, she would go to the marketplace and offer them a place to sleep and even feed them. Since Karlokeh was the gateway to Harper, people leaving various towns and villages would come to find transport vehicles to Pleebo, Harper, or sometimes as far as Monrovia. Mom would bring them to our house and provide for them with the little we had until they left. She was kind, gentle, and never emotional. She was, and still is, an angel to me. That was who the woman, my beloved mother, friend, and inspiration was.

CHAPTER 3

———•◆◆◆◆◆•———

Who Was Moffat Kateh-Mah Brown?

Moffat Kateh-Mah Brown

Moffat Kateh-Mah Brown was born in Takoradi, Ghana, on September 15, 1927, to the union of Sieh Kateh and Jugbeh Monoh.

Sieh Kateh was a master seafarer, always seeking opportunities for his tribesmen and women to explore greener pastures. Consequently, his tribe held him in high regard, even composing a song in his honor: "Sieh Kateh... Sieh Kateh jah nifa jolobo," meaning "Sieh Kateh, Sieh Kateh brought the Nifa tribe to civilization." Sieh Kateh was a tall man, standing about six feet tall and weighing between 255 to 285 pounds. His wife was around 4.9 to 5.0 feet tall and weighed approximately 130 pounds.

At ten years old, Kateh-Mah's mother sent him to Liberia to be raised by her sister, Tilly Monoh, to prepare him for his role as the Traditional High Priest of the Nifa Tribe. His uncle Wlubo, the current high priest, was to begin the initiation rituals. However, Kateh-Mah soon realized that becoming the successor was unlikely, as Wlubo was still young and energetic. Kateh-Mah continued his schooling but didn't go far because he got a woman pregnant and had to marry her according to custom. They had a boy who died a few months after birth, followed by three other infant boys who died mysteriously. This prompted Moffat Kateh-Mah Brown to decide that his best chance to make a positive contribution to society was through education and becoming a teacher.

Moffat enrolled at Waterside Elementary School completed elementary education, and then moved to Harper, where he entered Our Lady of Fatima High School. In the 10th grade, as an adult, he seized an opportunity when the County Education Officer recruited him and others to establish schools in rural areas. This was part of President William V. S. Tubman's initiative to open up rural areas and educate future leaders. Mr. Brown took up the challenge and was sent to the new, booming town of Karlokeh to establish a school. This school was named M. H. Gibson Elementary (grades 1-6) and later expanded to include a junior high (grades 1-9). It is now known as Karlokeh Elementary, Junior, and Senior High School.

Moffat ran from becoming a priest because it was clear that age wasn't on his side, and the mysterious deaths of his male children were traditionally linked to his father-in-law. He divorced his wife and moved to a new environment with a new zeal to make a difference. Meanwhile, Edith Wallace, devastated by her life's circumstances, decided to find her mother and start anew. While in Karlokeh, Kateh-Mah met Edith Hneyanneh Wallace. They fell in love, got married, and had two boys. The oldest boy was destined to be the traditional high priest to succeed his Grand Uncle.

Four days after my birth, my Grand Uncle, Wlubo, came to initiate me into the traditional priesthood. But he was disappointed. According to the story, he walked into the bedroom early that morning without speaking to anyone. As soon as he saw me, he realized that I was not fit for the role. Frustrated, he slammed the door and walked out, saying, "I expected to see a man, but the child in there is not prepared for this task. He is a white man, which means he will be educated and may not want to be part of the tradition." With the failure to initiate me, my formal education began early. I started learning the alphabet before my first birthday. Many people wondered why I was "rushing in life," as we say in Liberia.

Moffat Kateh-Mah Brown worked tirelessly to ensure that the school was opened and operational. After many years of running M. H. Gibson Elementary and Junior High School, it became crucial to elevate the school to a Senior High level. The surrounding towns and villages had kindergarten and elementary schools, but they needed a nearby high school for further education. Karlokeh, now a city hosting the offices of the Karluway District, saw the school expand to a senior high school.

CHAPTER 4

———·◆◆◆◆·———

My Desire to Be a Medical Doctor

Fulfilling the Role of the Traditional Priesthood Through Medicine

Although my family was disappointed that I didn't enter the priesthood of our Liberian tradition, I sometimes wonder if I fulfilled
that priestly role as a medical doctor. The distinguished African intel-
lectual John Mbiti wrote about priests and herbalists, noting, "Divin-
ers or medicine-men sometimes have religious functions[1]." Ozioma and
Chinwe supported Mbiti's description, explaining how, in the African
tradition, a "traditional healer" can also function as a high priest.[2] In a
sense, although I did not become the traditional high priest of our tribe,
my calling to the medical field served that purpose.

---

[1] John Mbiti. African Religions and Philosophy. (Oxford: Heinemann Educational Pub-
lications, 2nd edition, 1997) 68.
[2] Ozioma, Ezekwesili-Ofili, Nwamaka Chinwe, Okaka. "Herbal Medicines in African
Traditional Medicine" In Herbal Medicine, edited by Philip Builders. London: IntechOpen, 2019. 10.5772/intechopen.80348

Since Karlokeh had a government school, many kids from various towns and villages attended it when they were strong enough to walk the many hours from their homes. Consequently, it wasn't uncommon to have an eight- or ten-year-old child in kindergarten. I and a few other kids were different because my father was a teacher, and my mother was eager to have her son learn what she was deprived of. I started formal education as Francis Brown at the age of two, and by my fourth birthday, I was promoted to the First Grade. I was the youngest in the class, but interestingly, the school had first and second graders in the same classroom due to limited space and instructors. When I was in the first grade, I could do the classwork that the third graders were doing. It was bizarre because I was the youngest in the class. Whenever there was a spelling bee, I would go to the tail end of the line, knowing many of my classmates would miss the word, giving me the opportunity to spell it and beat them in their palms with a rattan. As a result, my classmates teased me, saying I was smart because my mother was a witch and had given me an extra eye; hence, I had three eyes.

I was affectionately called Boy Brown because my dad was Moffat Brown. I had so much admiration for my mother that we were like best friends. She never called me by my name but used a particular whistle or other affectionate names like White-Man, Eagle, One Who Never Gets Tired, Oldman (because I never wanted to play with my peers; all of my friends were older people), etc. My peers teased me about my mother's one eye. It bothered me, and I cried, becoming curious about why my mother had one eye. That didn't extinguish my desire to excel in school, where I was offered double promotions. My father, however, rejected these opportunities for me to begin a new grade mid-year. It was painful, and the other teachers were not happy, but they had to obey my father since he was the principal. Thus, I had to sit in the same classroom twice while topping the class.

Due to the incident after my sister's birthday celebration, my mother never celebrated birthdays for any of her kids. Instead, she would wake us up by 4 to 5 am, pray with us, and then wish us a Happy Birthday. That was a memorable tradition. On the morning of my twelfth birthday, my mother woke me up, prayed for me, and wished me a happy birthday. As she was about to leave the room, I asked, "Mom, can I ask you a question?"

She asked, "What is it?"

Curious as always, I said, "Mom," with tears running down my cheek, "Why do you have one eye?" She looked at me and began to weep, and I wept with her.

With her voice cracking, she said, "Son, when I was a child, I had an infection in my eye. There was no doctor in the village, so I was taken to an herbalist who put herbs in the eye, and I lost it. To console my guardian and me, the herbalist said that if he hadn't done it, I would have lost both eyes. As a result, I was teased by my peers and became so ashamed that I didn't go to school. This is why I prayed continuously to God to give you the wisdom and strength to study hard and get the education I was denied."

In consolation, I said, "Mom, when I grow up, I will be a doctor."

She responded, "Son, you have the ability to achieve anything positive you wish. The key is hard work, dedication, commitment, and trusting God; you will succeed."

That was the inspiration and motivation I needed to face the world. With those words from my mother, I decided to take practical steps. The first was to befriend the Physician Assistant (PA) in town. I would run errands and undertake other chores for him, like cooking his food and washing his clothes. As a result, I was introduced to minor medical activities, such as holding patients during injections, comforting young boys during circumcisions, sterilizing circumcision sets, preparing gauze, and other medical-related tasks. It was fun for me. After the tenure of one PA, I volunteered my services to the next and continued the same activities.

My curiosity in medicine was profound, and my relationship with the PA was valuable. One crucial morning, my sister and I went to school. While standing in line with other students to hoist the Liberian flag, I noticed a small stream of blood running down my sister's leg.

I immediately whispered, "Sister, there's blood running down your leg. Did you hurt yourself?"

She glanced at her legs and went to one of the female teachers, who assisted her and asked her to go home. The entire time I was in class, I focused on my sister, wanting to know what had happened to her. At lunchtime, I rushed home and called my mom with our usual whistle. She came out of the kitchen and said, "What are you doing here? You're supposed to be in school."

"It's lunchtime," I told her.

"So why are you here?" she asked again.

"Sister got cut this morning, and I wanted to know how she is doing. I didn't see her in school."

"No, your sister is in school," Mom told me.

"But Mom, what happened to her? Did she go to the clinic? I saw blood on her."

"Yes," Mom answered reluctantly.

I then returned to school. As I approached the campus, I saw the PA and asked if he had seen my sister. He said no. That bothered me greatly. After school, I saw my sister and asked her how she was doing. She said she was fine.

"Did you go to the clinic?" She refused to answer.

At home, I told my mom, "I saw the PA, and he said sister did not go to the clinic."

Realizing I wouldn't stop until I found out the truth, Mom informed me that my sister had become a woman. If she played around with men and became intimate and sexual, she could have a child. This answer raised more questions, but I left it alone.

Although I stayed busy assisting the PA and doing chores, my focus on school never ceased. I took the Ninth Grade National Examination while in eighth grade. I was offered double promotions several times, which my father rejected. However, this time was different. My class sponsor, the late teacher Augustine Davis, stood firm. He indicated that he knew the abilities of all his students and wasn't prepared to deviate from his plans. He submitted my name for the National Examination roster without further consultation with my father.

My curiosity led me to change my name from Francis Nah Brown to Francis Nah Kateh. I was born and raised in Karlokeh. When I entered the sixth grade, a relative visited and discussed our ethnic background. He explained how our tribe, Nifa, was sandwiched between the Garraway and Buah ethnic groups, predominantly "Grebo." During the discussion, he told me my grandfather, Sieh Kateh, was originally from Picnicess, near Barclayville, Grand Kru County. During wars between the Greboes and the Kru, the Nifa tribe often served as a reconnaissance group. My family were the custodians of the various gods protecting the tribe, making us the high priestly family.

After this discussion, I asked my dad why his name was Moffat Brown instead of Kateh. He explained that when he came from Ghana, where he was born, and since his father was often away, his father trusted a friend named Moffat Brown to serve as his guardian. So, he took his name. I respectfully told my dad that while it was good that he carried his guardian's name, I couldn't do the same for someone I didn't know. I changed my name to Francis Theophilus Nah Katison, believing it sounded more civilized and Anglicized. However, the full name couldn't fit on the National Examination form, so I shortened it to Francis Nah Kateh, which I carry today.

When I took the Ninth Grade National Examination in the eighth grade, my results were outstanding, breaking records at the school. Just before publishing this memoir, I learned my record still stands, with a cumulative average of 94.4%, placing sixth in the country at M. H. Gibson Elementary and Junior High School, now Karlokeh Junior and Senior High School.

My ambition was gradually cultivated, mostly inspired by my mother. After graduating from Junior High School, there was no Senior High School in Karlokeh, so I matriculated to Our Lady of Fatima High School in Harper, Maryland County. Living in Harper wasn't difficult because my father's younger brother, also a teacher, lived there, and my father built the house they lived in. A small room was made available for me.

CHAPTER 5

———— ◆◆◆◆ ————

Living in Harper

Living in Harper was a challenge, not because of housing, but because survival required navigating strict rules and harsh conditions. My uncle was quiet and dominated by his wife. To succeed, I had to quickly learn to obey her strict rules. Each morning, we had to fill a 52-gallon barrel with water from a nearby well, cook meals after school, and fetch firewood on Saturdays from wooded areas on the outskirts of Harper. Failing to meet her expectations made life miserable.

At the tender age of ten, my mother taught me how to cook and do household chores. Her friends would chide her, saying, "Why are you turning this young man into a female? You're teaching him to cook, clean, iron, and do laundry; these aren't men's work." But my mother persisted, ensuring I became independent. So, when my uncle's wife imposed these rules, it wasn't an issue for me, except for fetching firewood, which was always a long journey outside Harper.

On Saturdays, the streets were crowded with young people from various parts of Maryland and surrounding counties, all living with different families and gathering firewood. We obtained dried firewood from behind Bishop Ferguson High School Campus and along the road to Fishtown, a few miles outside Harper. The wood was primarily used for cooking and baking. At my uncle's residence, it was mandatory for all children of extended relatives to gather firewood, while the biological children were exempt. The journey to find good firewood was long, about ten to fifteen kilometers. We would gather large bundles, place them on our heads, and return home. I often experienced severe headaches, particularly at the back of my head, which I later learned was the occipital region. This area controls vision, and my visual problems were becoming an issue.

My inability to fetch wood was sometimes unacceptable to my uncle's wife. For peace, I compromised occasionally, but when I couldn't gather firewood on Saturdays, I faced starvation. If I didn't fetch firewood, I wasn't allowed to cook. If Palm Butter was the soup for the day, my uncle's wife would give me a few uncooked palm nuts and some raw rice as my meal. On the days I was allowed to cook, my routine was to leave school by 1:35 PM, get home by 2:00 PM, walk to the market, buy ingredients, and start cooking by 3:00 or 3:30 PM. By the time the food was ready, it would be about 6:00 or 6:30 PM. My uncle's wife would then distribute the food in various bowls, and we would eat and then study before bed.

One unforgettable incident happened on a hot Thursday afternoon. While leaving the market with a wheelbarrow full of cooking ingredients, someone called my name. I turned around but saw no one. Suddenly, a police officer named Sgt. Moses slapped me in my left ear, shouting, "You grona boy!" ("Grona" was local slang for a useless person.) "You want to hit me with the wheelbarrow?" Since I often had no food until the end of the day, the slap caused me to faint. When I regained consciousness, people were sprinkling cold water on me. Without a word, I pushed the wheelbarrow home, cooked, ate, and then went to the police station to file a complaint.

At the station, I reported the incident to the commander, Benedictus. As I was concluding my statement, Officer Moses arrived and continued to call me "grona." I responded, "Keep calling me grona, but when I'm out of college, I'll charge you for slapping me without reason, and you'll spend the rest of your life in jail."

The commander told me I was being disrespectful and threatened to detain me if I uttered another word. I left the station with tears running down my cheeks, feeling a mix of anger and determination. On my way home, I pondered how I would seek revenge in the future when I became educated and influential. I imagined citing Officer Moses to my office, explaining how his actions had motivated me, and giving him a choice: either confess on the radio or face a 24-hour detention.

Living in a community where human rights were not prioritized, I often felt powerless against those who abused their authority. However, I remained conscious of the injustices around me and dreamed of a future where I could make a difference and protect the rights of others.

CHAPTER 6

A Student at Our Lady of Fatima High School

Upon my graduation from Junior High School, my dad asked me to enroll at Our Lady of Fatima High School, assuring me that he would find a way to pay my tuition. I took the admission test and was successful. When school officially opened, I began classes at the high school. On my second day, the principal called me to her office. Fear gripped me because, from what I gathered from other students, the principal rarely called students to her office unless they were in trouble.

When I got to the office, the principal asked, "Can you tell time?" I answered yes. She showed me her wristwatch and asked, "What time is it?" I called the time, then she pointed to the clock in her office and repeated the question. I gave her the time again, and she said, "Henceforth, you are the timekeeper of the school."

Every morning, I got up by 4:30 AM to fetch water before going to school. I filled a large empty barrel with water and made sure I got to campus before 7:05 AM to carry out my duties as the timekeeper.

Anyone advancing educationally in Harper needed an unbreakable commitment to education. Living there was challenging, but through God's grace, I succeeded and graduated from high school. My journey, however, was not without its challenges. One of the most painful experiences that could have interrupted my secondary education occurred when my first cousin and her mother destroyed all my textbooks and clothes, with her mother, my uncle's wife supporting her actions.

It happened on a fateful Tuesday after school. I came home and found my uncle's wife's sister's infant son crying. I picked him up to comfort him, but his diaper was filled with feces, which spilled onto my shirt.

I said to my cousin, "Why didn't you warn me that the child had diarrhea? Now my shirt is a mess."

She became belligerent and said, "If you say a word, I'll take off his diaper and throw it at you."

I challenged her, "You can't!"

She immediately took off the soiled diaper and threw it at me as I tried to escape. The diaper landed on my back, covering my shirt in filth. I became angry and hit her, then went to my room. The next-door neighbor, seeing what had happened, told me to leave her alone and wash my shirt. While I was washing the shirt, my cousin took a machete and threatened to chop me up. The neighbor, knowing what my cousin's reaction would be, asked me to go to her house. While I was there, my cousin broke into my room, collected my clothes and books, and threw them into the nearby river basin.

When my uncle's wife, Aunt Esther, came home, the neighbor took me back to the house, and I explained what had happened. The neighbor, who witnessed everything, supported my account. Aunt Esther's response was chilling: "He is lucky. If I had been around, I would have told her to get gasoline and set everything ablaze."

The neighbor asked, "Why? What has this young man done to you?"

I began to cry and decided that was it. No more school. I was tired of the painful and brutal experiences just to get an education. Enough was enough. I was going back to the village to help my mom on the farm.

When I returned to the village late that evening, my mom had just come back from the farm. She was frightened and wanted to know what had happened. I explained everything, and we both cried before going to bed.

Early the next morning, Mom woke me up and said, "You will stay home today, but you and your dad will return to Harper tomorrow. We will replace some of your clothes, and your dad will talk to the school to see if they can help replace your books. My son, listen to me; nothing will stop you from being educated. The book I don't know, you will learn it all. You must go back to school. I am not educated; this is why I go to the farm daily to ensure that you all succeed."

I was angry and felt I couldn't bear the difficulties of living with my uncle any longer. Despite my reasons, my mother was determined for me to return and continue my education. My dad and I returned, and the case was investigated. During the investigation, my uncle's wife remained adamant, repeating her statement about setting my things ablaze. At the end of the investigation, my father told his younger brother that he would have to leave the house due to his wife's wickedness and disrespect. Others intervened and apologized, resolving the matter. It was an eye-opener. I continued to be obedient, did my chores, and avoided any further conflicts with my cousin until my graduation.

December, 1982 13 graduates out of 21 in the class with 8 passing all of the courses of the National Exams and 5 repeating one course each

Upon my graduation, the family wanted to have a ceremony, but I refused. I wasn't prepared to pretend and make those who had been so wicked to me happy. I was grateful to God for giving me the strength and endurance to complete high school. I told them I was going back to the village to spend time with my mom. It was unusual for students at that time to graduate without a big party where friends and foes would gather to celebrate, but that wasn't me. My life was geared toward the ultimate goal of becoming a doctor.

CHAPTER 7

————— ◆◆◆◆◆ —————

Matriculation After High School

Before graduation, my father asked me what I wanted to be, aside from the doctor I always talked about with my mom. He said, "Have you seen anyone from our tribe or community become a doctor?" In a quiet tone, I replied, "Daddy, I've always told you that I want to be a doctor."

My father yelled, "It's not possible. You will become a teacher." He then took the liberty of registering me for the placement exam to attend ZRTTI (Zorzor Rural Teacher Training Institute) in north-central Liberia, a few hundred miles from Maryland County. Knowing my father's temperament, I didn't utter a word but went with the flow. I had an alternative plan.

When the date for the teacher training school exam came, I went to Pleebo City, where the exam was to be administered, but I didn't show up at the Examination Hall. After a couple of weeks, the results arrived. My father asked me about the outcome because he had heard the results were out. I told him I was very sorry and afraid to tell him that I had flunked the exams.

A proverb says, "A child can run but cannot hide." My father, being a school principal, had access to the County and District Education Officers through whom the results were sent. He met with the District Education Officer a few weeks later and expressed his shock that I didn't pass

27

the admission test. The officer said that couldn't be true because, based on my performance on the ninth-grade examination, he couldn't believe I didn't pass. He then pulled out the results and saw that I didn't show up for the exam.

My father returned late that afternoon. From the way he was walking toward the house, I knew something terrible had happened. I quickly hid myself. When a child does something positive, the father usually takes the glory, but when it's bad, it's blamed on the mother.

When my father got to the house, he shouted for my mother, "Edith! Edith! Where are you? I told you about your son, and you keep encouraging him to be mischievous. Now, he has brought disgrace to me. He couldn't take the test because he felt unprepared, so he didn't show up? I knew he was up to something when he lied and said he didn't pass. Where is he?"

Mom pleaded, "I beg you, for God's sake, please let me talk to him. I'm sure he had a reason, and we will find a way around it."

I heard every word from my hideout, realizing I couldn't easily deceive my father. However, my plan was accomplished because now I would have to wait a year before taking the examination again.

Due to the tension between my father and me, I spent most of my time with my mother on the farm. I helped her pick weeds from among the rice and plant peppers, cassava, and more.

During our time together on the farm, I kept telling her that I wanted to become a doctor. Her response was always, "My son, whatever you want to do in life, if you are committed and put your trust in God, it will come to pass."

After several months of my dream seeming out of reach, frustration set in, and I began to doubt if it would ever come true. During that period of ambivalence, I told my mother I had decided to write a letter to the then-Bishop of the United Methodist Church, Rev. Dr. Arthur F. Kulah, to allow me to volunteer at the Ganta United Methodist Hospital in Ganta, Nimba County. She encouraged me to write the letter, which I mailed from Harper. A few weeks later, I received a response from the bishop approving my request and enclosing twenty-five United States dollars wrapped in carbon paper for transportation from Karlokeh to Ganta. It was an exciting moment for me. I shared the letter with my mother, who was so grateful and began praising God that my dream was gradually becoming a reality.

My younger brother, Anthony Sieh Kateh, was so happy that he called me "Bofugunku," one of my childhood names. As a teenager, I often participated in programs, reciting speeches. One of those recitations from when I was five, which I still remember, goes like this:

"I would like to introduce myself to you all tonight. My name is Bofugunku. I was born in Casablanca, Morocco. I have traveled for forty years, forty months, forty weeks, forty days, forty hours, forty minutes, and forty seconds to inform the world that a hint to a wise is quite sufficient, but to a hidden it is dangerous... I thank you."

He said, "I know you will be a doctor, and when you become one, you will never work on me." I laughed, but our mom wasn't happy about his remark. She said, "Stop saying that. When your brother becomes a doctor, he will take care of all of us. Would you feel good going to someone else?" My brother then said, "Big Brother, I was just joking."

Later that evening, my mom informed my dad that I had received a response from Bishop Kulah regarding my willingness to volunteer at Ganta United Methodist Hospital. He downplayed the information, but a few days later, it was time to say goodbye.

CHAPTER 8

---◆◆◆---

Moving to Ganta

On that providential morning, my mom got up at 4:00 AM, worshipping and expressing gratitude to God. I was finally beginning the first step toward what I had always wanted to do since I was twelve years old: embark on my career pathway in Ganta. Along with the letter from Bishop Kulah was another letter for Dr. Augustine Tetteh, the Chief Medical Officer of the Ganta United Methodist Hospital.

After spending over sixteen hours riding in a public commercial vehicle, I finally arrived in Ganta. I went directly to Dr. Augustine Tetteh and handed him the letter from the bishop. Dr. Tetteh welcomed me warmly and directed me to Rev. Juwle, the Station Manager of the Ganta United Methodist Mission, to find me a place to live. I was given a self-contained apartment attached to Rev. Juwle's house.

Without any professional background, I was assigned various duties such as cleaning the floors, calling the names of patients, and sometimes helping in the kitchen. While working at the hospital, my desire to become a doctor only intensified.

Due to my growing interest in medicine, I was always curious about what was happening in the operating room. Whenever I was free, I would peep through the glass door to observe the procedures. One day, Dr. Walter

Stephenson, a missionary general surgeon and former military doctor at the hospital, asked me, "Why do you peep through the glass door when we are doing surgery?"

"I want to become a doctor," I told him.

"A doctor?" he asked, surprised.

"Yes," I replied.

He quickly interrupted, "You are a cleaner!"

"Yes, I am," I said. "But I am volunteering."

"Wow!" Dr. Stephenson's tone changed to one of acceptance. "That is good to know. Are you in or out of high school?"

"I've graduated," I told him.

"But why are you not in college if you are out of high school?" he asked.

"My parents can't afford to send me to college."

"Where are your parents?"

Dr. Stephenson took me to his office, and I explained what motivated me to become a doctor.

"That is inspiring," he said. "Great! It is good to know."

The next morning, he called me into his office and asked me to come into the operating room to observe a surgery. It was a C-section (removal of a baby from the womb surgically).

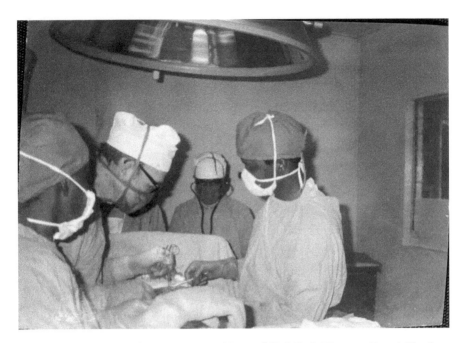

The first observation of surgery at Ganta Hospital (R-L Dr Sei Parwon, Francis Kateh, Garrison Kerwilliam, Nurse Anesthetist (shoulder only), Dr. Walter Stephenson & Mom Martha, Scrub Nurse)

"Please check when the University of Liberia will administer its next admission exams, and I will make the payment. When you take the exams and pass, I will support you." That was the day my path to my dream became clearer. I went to my room and thanked God for the intervention that would happen if I passed the admission test. The next day, Dr. Stephenson gave me funds for transportation, and I went to Monrovia to register for the admission test.

The registration process took three days, and not understanding how it worked, I followed all the procedures. It was a nightmare, but I eventually made it. The morning I went to get my admission test number was when students of the University of Liberia began demonstrating against President Samuel Doe. I found myself behind the TH Building (Tubman Hall). I quickly ran behind the University Clinic building, only to face a soldier. He apparently wanted to shoot, but whether the shot got jammed or he ran out of ammunition, he immediately turned and hit me with part of his gun. I didn't stop running and found myself down the Jallah's Town Hill. A driver stopped for me, and I got in his vehicle. When we got to Broad Street, I began to feel a tingling sensation in my lower extremities. I was quickly rushed to Snapper Hill Clinic and later to Ganta Hospital, where I made a full recovery.

I took the entrance examination at the University of Liberia and passed. It was a joyful occasion for me, and I quickly sent a letter to the village informing my parents that I had finally accomplished the first major step towards becoming a doctor. I returned to Ganta and was given funds to travel to Karlokeh to see my parents. My mother was so proud that I had begun the first step of my journey and was grateful to God for leading me towards my ambition.

After spending approximately two weeks with my parents, my dad called a distant relative to arrange for me to live with him in Monrovia. I left Karlokeh and went to live in New Kru Town, Bushrod Island, Montserrado County. I lived with my distant relative, who served as my guardian because my father was afraid I couldn't live alone. That was an interesting experience.

My "uncle," Mr. Alfred Gbe, and his wife received me warmly. Unfortunately, he didn't have a place of his own but rented a single room in a house with other renters. He had four kids, and at night, they all slept on the floor before the bed. During my first week, I slept on the floor with the kids, who all peed on the mat, soaking it with urine. The owner of the house called me one morning and explained that she had spoken with her nephew so I could sleep on a mattress in the living room. That was a relief, and I was grateful to her and the family. I slept on the mattress for almost six months before finally getting my own room.

About a year into the university, my sponsor returned to the United States due to illness and, unfortunately, did not recover. Life became difficult for me. Four times, I had to walk from New Kru Town to the University of Liberia main campus, about 9.8 miles away, because I couldn't afford the twenty-five cents for the bus. It was interesting that I met a few others going through the same trials. One such friend was the late former Minister of Public Works and Distinguished Senator from Sinoe County, Mobutu Vlah Nyepan. Despite our challenges, he was always neatly dressed, outspoken, and willing to assist if he could. I depended on an arrangement where I would pass by his girlfriend's house every morning to get a portion of her fund for lunch to take to him. He would then share his portion with me, which assisted me greatly with transportation and food.

Things changed for the better one day. It was late afternoon, and I had only twenty-five cents for the bus home, but I was very hungry. I decided to stop at a Fula shop, a small dry goods shop just below the University of Liberia campus, where I bought what we called "Fula bread." It was a thin, long, round bread cut open in half and spread with mayonnaise. I completed the order by adding a hard-boiled egg and a cold bottle of Coca-Cola. At the time, Coca-Cola had a promotion with cash prizes under bottle caps. When I opened the bottle, I threw the cap on the floor. The little girl helping in the shop told me to pick up the cap to see if I had won. I was hesitant, but she insisted until I reluctantly picked it up. I discovered I had just won thirty-five US dollars. Excited, I gave the girl five dollars and kept thirty. With the money, I took a taxi home, bought a fifty-pound bag of rice, and had enough left for soup ingredients and transportation.

College life in the city was tough, especially for those of us from counties outside Monrovia. My faith in God grew as I saw how He always provided relief. We were committed, dedicated, and prepared to do everything in our power to achieve our goal of acquiring a quality education.

CHAPTER 9

Aunt Mary Tweh Workoleh Sepeah

Aunt Mary Workoleh Sepeah and Grand Kids

My Aunt Workoleh, my dad's first cousin, played a vital role in my life by providing food and other necessities. She was a businesswoman selling cooked food at the Harbel General Market and was very supportive of me. Aunt Workoleh was married traditionally to the late Mr. Sepeah, an employee of Firestone Rubber Company, where he worked as a foreman until his retirement. Before his retirement, Aunt Workoleh had saved enough money from her business to build a modern brick house in Dolo's Town, Margibi County, just outside Harbel and near the Roberts International Airport (RIA).

My second visit to Monrovia in 1981 was based on an invitation from Aunt Workorleh, who paid for my trip from Maryland County to Harbel Firestone via Monrovia. When I enrolled at the University of Liberia, I often traveled to Dolo's Town during holidays or weekends to help her prepare the foods she sold in the market. When I returned to school, she would prepare various soups for me and give me some money to keep me going. Unfortunately, I didn't have a refrigerator, so I had to heat up the soups every morning to keep them from spoiling.

I was always grateful to my aunt for filling the physical void of hunger for a few days at a time, giving me the energy needed to study with the hope that one day, all would be well. Building on my mother's belief that if you work hard, stay committed, make sacrifices, and have faith in God, you will succeed, I remained positive and grateful for any assistance I received from others.

During one of my visits to Dolo's Town, a thief broke into my room and stole my mattress, clothes, and other belongings. It was devastating. To seek refuge, I turned to a friend whom I considered my "play-father"—an expression used to describe an older friend who cares for and takes responsibility for a younger one. Bolton Tarleh Nyema was that person. Mr. Nyema was the first President of the United Methodist Young Adult Fellowship of the Liberia Annual Conference, and I served as the Assistant Secretary General. We were all struggling as young people from out of town without primary relatives in the city. Mr. Nyema, who taught evening classes, allowed me to stay with him for over five months.

It still amazes me, but I have come to realize that God provides when one depends on Him. He uses others to come to the rescue in times of need, helping to fulfill what He has destined for that person. During my educational journey, while seeking refuge at Mr. Nyema's place, I decided to spend my vacation at the Ganta United Methodist Hospital. At Ganta, a missionary named Lois Kohler from Albert Lea, Minnesota, decided to assist me financially. This was a miracle, and with additional financial assistance from Mr. Nyema and others, I was able to find a place to live near the University of Liberia campus.

When I completed most of my general courses on the main campus of the University, I had to relocate to the Science College, located in Fendell, Montserrado County, more than 15 miles from the main campus.

CHAPTER 10

————— ◆◆◆◆◆ —————

Fendell: The Science College

M oving to Glabo's Town, Lower Careysburg, Montserrado County, near the Science College, was a lifesaver. This well-calculated decision yielded great dividends. It cut down so many expenses since it was within walking distance to my classes. Living in Glabo's Town was miraculous because many great things occurred. I was elected General Secretary of the United Methodist Young Adult Fellowship, witnessed a real-life miracle, and received the scholarship that took me to the United States.

In Glabo's Town, I lived in a house with two others, Mr. Dominic Reenie and Gibson. They shared a room, while I initially had the resources to pay for my own. When Lois Kohler, the missionary, left, life became a bit challenging, but I wasn't deterred because I knew there would always be a miracle. It happened one auspicious Friday afternoon when the landlord, Mr. Badio, sent a message that we needed to pay rent since we were months behind. Dominic and Gibson went to Monrovia after class, but I had no funds for transportation, so I stayed behind.

I walked to the central attraction in the town, a shop owned by Mr. Warner. While sitting there, a pickup truck sped by, and a bag fell from the back with a loud clanging sound. The pickup stopped, and two people got out, quickly picked up what they could, and drove off. Those of us at the shop ran to the spot where the bag had fallen and picked up as many five-dollar coins as we could find. I collected thirty-five coins, amounting to one hundred and seventy-five dollars. That amount paid off my debt and covered three additional months of rent. I also bought a half-bag of rice and other necessities, allowing me to accommodate my late mother during her visit to Glabo's Town—the only trip she ever made to Monrovia. After paying my rent, I could study hard and have friends over to study or spend weekends.

It was in Glabo's Town that a significant decision about leadership within the Young Adult Fellowship was made. When Bolton Tarleh Nyema, the Godfather of the Dynasty—a group committed to supporting Bishop Arthur F. Kulah—left the helm of the Young Adult Fellowship, it was decided that leadership would follow a succession plan. Brother Francis K. Zayzay, the Vice President, was to succeed him. I was to succeed Brother G. Roosevelt Tule as General Secretary, and Tule would become the First Vice President. However, Brother Tule decided to run for President of the Young Adult Fellowship, citing his popularity within the Monrovia and St. Paul River Districts. We mobilized support from other districts to back Brother Zayzay's bid for President.

When the delegations assembled at the convention site, Brother Tule, realizing his slim chances of winning, approached me and said, "Francis, you will remain as Assistant Secretary-General because I am going to retain my position." I didn't respond, knowing it wouldn't be possible. On the day of the election, Brother Francis Zayzay won the presidency, defeating Brother Arthur Saywala, and I defeated Brother Roosevelt Tule three to one as General Secretary.

One Friday afternoon, my boss, President of the Young Adult Conference Brother Francis Zayzay, informed me that a team was coming to see me in Glabo's Town. That Saturday morning, I welcomed Brothers Sundah, Brewer, Zayzay, Toe, and Kelly for a major discussion about the upcoming election. Brother Edward Brewer, the conference Treasurer, couldn't become President due to the ascendency clause instilled by the Godfather. Brother Zayzay did his best to convince Brother Brewer not to contest for the Presidency, arguing it was my turn. The discussion went on for hours, but I insisted on running. In the end, because of a divine plan for my life, that battle was never fought. Brother Brewer won the election because I had to leave, thus keeping the Dynasty in power. I had received a scholarship to study in the United States, so I didn't pursue any actions that could have caused division within the Dynasty. It all worked out in favor of the Dynasty.

CHAPTER 11

———◆◆◆◆◆◆·———

The Turnaround

On a gloomy day, I left Glabo's Town and went to Monrovia to visit Bishop Kulah. Upon arrival at the bishop's office, his secretary asked why I looked so sad. With tears running down my cheeks, I explained that the university had shut down again, delaying the completion of my studies.

She said, "Go and see your father, the bishop."

"Why?" I asked.

She responded quietly, "Please, go and see the bishop."

I approached the bishop's office door and knocked softly.

"Who is it?" the bishop asked.

"It is your son," I responded.

"Come in," I heard him say, and I gently walked in.

"Wow! Why are you looking so sad?" the bishop asked.

"The university has been shut down again," I told him.

"Oh, I just heard it on the radio. What is your level now at the university?"

"A junior student majoring in Zoology and minoring in Chemistry."

"Okay... Did you and Jerry get the scholarship to Bishop College in Texas?"

"No, Sir," I said. "After the vetting process and submission of the required documents, I haven't heard from them."

"Don't worry." He then pulled open a side drawer and called his office assistant, Stephen Blawah. "Where is the letter I received the other day?" he asked Stephen.

"I have it here, Bishop," Stephen answered.

"Please bring it."

Stephen brought the letter, and then the bishop called his secretary, "Ora, please come. Do a communication to MacMurray College recommending this young man. He deserves it." She smiled in agreement and prepared the letter immediately for his signature.

Several weeks later, the university reopened. My study mates and I were preparing for an Organic Chemistry exam when Stephen came over that evening, called me aside, and said, "I have your I-20 Form. You have been approved to travel to the United States to continue your education." It was the most joyous day of my life. I immediately ran back into the room where my colleagues were studying and announced that I had just received my I-20 Form, enabling me to get a visa for the United States, where a scholarship had been awarded to me to pursue a bachelor's degree in biology. Therefore, I was leaving and wouldn't be taking the exam the next day. I gave away my study guide for Organic Chemistry. One of my friends said, "I will keep it for you whenever you return."

I returned to my room, filled with joy and praise for God's continuous guidance, but I was also worried about my mother and how I would come up with the plane ticket to travel to the United States. The church didn't have the funds to cover it. I went through the various procedures and processes and finally got the visa. After contemplating and pondering a way forward, I raised some money to transport myself back to Karlokeh, Maryland County, Liberia, to bid my parents goodbye and to see if they could help raise funds for my ticket.

I traveled to Karlokeh on a truck. The trip was long, but I couldn't wait to share the news with my mother. After traveling for more than twenty-four hours, I arrived, and my mother made some food for me. After taking a bath to wash off the road dust, I went to bed in my mother's room as usual.

At about three o'clock in the morning, my mother woke me up and asked, "Why are you here? School is still in session. I hope you didn't do something horrible."

"No, Mom," I responded. "I got a scholarship," and I pulled out my passport that contained the visa to the United States to continue my studies. She immediately began to sing and praise God for the blessings.

"I have always told you that if you are honest, committed, respectful, disciplined, and willing to go the extra mile, God will always bless you. I am so grateful to God for answering my prayers. Please, continue to do your best and don't bring disgrace to us. I am sure you are very respectful and committed, but you are going to a different country. Don't be carried away. God has given you an opportunity; please make the most of it."

In response, I began to shed tears and said, "Mom, I always told you I want to be a doctor, and God has heard our prayers. I will do all I can to reach my goal. It is an opportunity I will never abuse but make the most of."

The next day, my father arrived from Harper, where he had attended a workshop as the principal of the school in Karlokeh. He was furious when he saw me. "What are you doing here? You were supposed to be in school. I hope you didn't do something bad for which you may be suspended or running from school."

In her quiet tone, Mom said, "Please take a deep breath, and I will explain to you."

"No! You need to tell me now," my father demanded. "You have continuously protected your son to the extent that he refused my orders to go to a Teacher's College. Now it seems he can't make it in university, so he is back home bringing disgrace to the family."

"This is not the case. He is going to make us proud," she responded. As he continued to be upset, she called me and said, "Go get your passport and bring it." I did as she asked, and she showed it to him, saying, "God has answered our prayers. Our son will be going to the United States."

"My son?" my father said as if speaking to himself and to me.

"Yes," I responded.

Mom spoke with confidence that felt like she was chiding him a bit. "Your son, whom you always belittled."

Dad quietly entered the house and called both my mom and me. He said, "I am grateful to God and very proud of our son. I am sure he will bring pride to the family."

After spending a week in Karlokeh, I decided to return to Monrovia. I had written to the senior pastor and some prominent members of the S. Trowen Nagbe United Methodist Church to ask for assistance with funds to purchase a one-way ticket. My parents, not having much financially, blessed me and placed me in the care of our Lord and Savior, praying for favor upon my life. They gave me the little funds they had and added a large billy goat to be sold, with the proceeds added to whatever contributions I received to get the ticket.

Upon arrival in Monrovia, I went to the church and met the senior pastor. He told me the church didn't have extra funds, and he wasn't convinced the scholarship was a full scholarship. Disappointed and in despair, I continued to ask others for help. I even wrote to the administrator of Ganta United Methodist Hospital for assistance. None of my communications yielded positive responses. With about four days left before departure, I began to lose hope.

It was a day before my departure, and I still had no ticket. I went to the Methodist compound to see if someone from the church office could help. I met with Bro. Winston Smith, Treasurer, and he said it wasn't possible because the church was indebted to the Sahara Travel Agency, which usually issued tickets for conference travels. As I walked back to the main road, the Ganta Hospital administrator saw me and stopped.

"I thought you had left," he said. I began to cry. "No, I haven't because I can't get funds to pay for the ticket."

"I thought you wrote to Ganta Hospital," he said.

"Yes, I did, but I haven't gotten a response. If I don't leave by tomorrow, I may forfeit the process for the semester."

"Wow, come with me to Bro. Smith's office." We walked back to the office, and he instructed Mr. Smith to get the ticket and bill Ganta Hospital. Instantaneously, a call was made, and Santos, the driver, went to the travel agency and got the ticket. What a miracle!

CHAPTER 12

---◆◆◆◆◆◆---

The Trip to MacMurray College, Jacksonville, Illinois

Francis Nah Kateh, Jan. 19, 1989, two weeks prior to my departure
for MacMurray College, Jacksonville, IL

On the morning of February 1, 1989, with my ticket and visa in hand, I departed from Monrovia and went to Dolo's Town near Harbel, Firestone, to pack my things for the United States. A couple of friends arranged for the Youth Department bus of the Annual Conference to come and bid farewell to their General Secretary of the Methodist Young Adult Fellowship. It was a historic but sober moment.

After going through the ticketing process and bidding farewell to those who escorted me to the airport, something peculiar occurred. As I walked towards the KLM flight, I heard a voice telling me that I would never see my mother again. The voice was very clear as I boarded the Boeing 747 and took my seat. The flight was excellent, and I arrived in Chicago the next day after a layover and flight change at Schiphol International Terminal in Amsterdam. The only contact I had in the United States was Dr. Jim Goulding, the intermediary between the United Methodist Church in Liberia and the Central Illinois Conference, which had committed to sponsoring a Liberian student due to a request from Bishop Arthur F. Kulah.

When the pilot announced our descent to O'Hare International Airport in Chicago, Illinois, a song came on the speakers overhead, "The Living Years," written by Mike Rutherford and Christopher Neil, released on October 28, 1988. The lyrics of the song epitomized my journey to this new world and new culture. I felt like I was on a trampoline, jumping to change the future of my family, tribe, and country, provided I internalized the words of "The Living Years." The song reflected my relationship with my father but also made me feel a strong connection to my mother.

Every generation blames the one before
And all of their frustrations come beating on your door.
I know that I'm a prisoner to all my father held so dear.
I know that I'm a hostage to all his hopes and fears.
I just wish I could have told him in the living years

So, we open up a quarrel between the present and the past
We only sacrifice the future; it's the bitterness that lasts.

So, don't yield to the fortunes you sometimes see as fate.
It may have a new perspective on a different day.
And if you don't give up and don't give in, you may just be okay

So, say it, say it, say it loud, say it clear (oh, say it clear)
You can listen as well as you hear.
Because it's too late, it's too late (it's too late)
When we die (oh, when we die)
To admit we don't see eye to eye[3]

On arrival at Chicago O'Hare International Airport, the temperature was minus two degrees Fahrenheit with a wind chill factor of minus twenty degrees. It was the coldest I had ever felt. Coming from a tropical area with temperatures in the eighties, it felt like stepping from a furnace into a freezer. My entire body felt frozen. I couldn't feel my lips or my fingers. I had left Liberia without proper clothing for the cold Illinois weather, wearing only jeans, a long-sleeve shirt, and a sweater, with nothing to protect my head and hands. What made it worse was that I had to get off the flight and board a shuttle bus to the arrival gate. I almost refused to leave the KLM flight when the Windy City breeze hit my face. My biggest fear was who would meet me at the airport.

I braved the cold, got on the shuttle, and went through immigration, where I answered the appropriate questions and had my passport stamped. With a smile, the female immigration officer said, "Welcome to America." Those three words were so inspiring and rejuvenating. As I walked out of the immigration area, someone called my first name. I asked, "Dr. Goulding?" because he was the only contact, I had been in communication with. I was so grateful to Dr. Goulding for making the impossible possible by getting me to the United States and meeting me at the airport.

---

[3] Mike and the Mechanics. 1988. The Living Years

CHAPTER 13

———••◆••·———

Dr. Goulding's Narrative

**Dr. Jim Goulding handing me the Distinguished Alumni Award in 2001,
Ten yrs. after my graduation from MacMurray College**

"Dr. Francis Nah Kateh, MacMurray Class of 1991, the MacMurray College Alumni Association, on behalf of MacMurray College, is pleased to acknowledge your dedication and service to God, the United

Methodist Church, and your native country of Liberia by awarding you the Young Alumni Award for 2001." These words still echo in my mind as I recall the award ceremony in 2001. In reading my summary journey as the honoree, Dr. Goulding explained the following:

"In February 1989, Francis N. Kateh of Monrovia, Liberia, arrived at O'Hare International Airport in the midst of a snowstorm and sub-zero temperatures. I picked him up and put him on the train to Springfield, where he was met by a member of the Admission Staff. Coming from a tropical country, Francis did not have any warm clothing. If my memory is correct, a trip to Walmart took care of that need. Francis was the first of our four Liberian students to come to MacMurray College as a result of a partnership with the former Central Illinois Conference of the United Methodist Church (now part of the Illinois Great Rivers Annual Conference), whereby MacMurray College provided a full tuition scholarship for up to four years, and the churches of the Central Illinois Conference provided money for room and board, health insurance, and personal expenses.

"Francis graduated from MacMurray in 1991 with a bachelor's degree in biology. He went on to graduate from Spartan Health Science University School of Medicine in St. Lucia, West Indies, in December 1995 and did clinical work at Norwegian American Hospital in Chicago and an internship at John F. Kennedy Hospital in war-torn Monrovia, Liberia.

"Churches in the Central Illinois Conference paid for Francis' medical education through an Advance Special. And somehow, I became the manager of those finances. Francis has never told me how I got this honor. There were many times when the funds were nearly exhausted, and I would call Francis to tell him how low the funds were. This did not stop Francis. He had an abiding faith that he was called by God to become a medical doctor and help meet the needs of his people, who had suffered so much through years of civil war. He had confidence that the money would be there for him to complete his education. And it was.

"While Francis was in Monrovia, working at the JFK Hospital, fighters raided the hospital, and one grabbed him, stole his watch and other personal items, and tried to kill him. Kateh grabbed the fighter's hand, which was holding a knife, and held on until another fighter intervened, pointing out that he was a doctor and needed to stay alive.

"In an interview with the Illinois Great Rivers Conference Current in 1997, Dr. Kateh, commenting on this instance, said: 'God must have wanted me to stay there and care for the people. Out of that experience, I learned to be strong and to know without a doubt where God wanted me to be. God did not see me through school and bring me back here to be killed in less than a year.'

"After completing his intern year in Monrovia, Dr. Kateh joined the staff of the United Methodist Ganta Mission Hospital, where he was one of two licensed physicians on a staff that needed eight doctors. The hospital had been looted by warring factions but continued to treat 100-125 patients daily with a skeletal staff and little equipment or supplies. Dr. Kateh was made chief medical officer of the Ganta Mission Hospital in 1998. With help from the General Board of Global Ministry of the United Methodist Church, Dr. Kateh returned to the United States in 1997, 1999, and again in 2000 to appeal for funds and supplies for his hospital.

"Dr. Kateh possesses several admirable traits: a dynamic faith that God has called him to serve his native country as a medical doctor, patience to endure many years away from his family in years of study to become a medical doctor, discipline to organize his life and study time so that he could gain the necessary knowledge and skills to carry out God's purpose for his life, and a magnetic personality that won over people to support his education and now to support Ganta United Methodist Mission Hospital. You never say no to Francis Kateh.

"Francis Nah Kateh, you have brought honor to your alma mater through your commitment and sacrifice to meet the medical needs of the citizens of Liberia. It is with great pride that we recognize you as the recipient of the Young Alumni Award and present to you this plaque at MacMurray's Homecoming, October 13, 2001."

~ Written and presented by Dr. Jim Goulding,
Dean of the College

In recollection, Dr. Goulding took me from the airport to the train station, where he bought lunch for me, and I had my first cheeseburger. I got on the train for Springfield, Illinois, where a counselor from the admission department awaited my arrival. What was amazing was that the counselor met me with a winter jacket and a pair of gloves. That was excellent. The trip from Springfield to Jacksonville, IL, was smooth but full of ambivalence. There were so many questions on my mind about this new life and opportunity that God had given me. Looking out the train window, I saw flat land for miles with snow covering the ground. Coming from a country with diverse topography, from tropical rainforests to savannas and beautiful beaches along the Atlantic Ocean, I wondered what this new adventure would be like. Will I be able to survive? How will I be accepted in this new country, city, and college? Will I be treated differently because of my color, accent, or height? Despite the many questions about myself and my personal journey, I had other questions about my expectations at MacMurray, which were soon answered. I hoped everyone would embody the warmth and kindness of Dr. Goulding, the counselor, and his fiancée. After an hour-plus drive from the train station, I arrived on campus and was immediately taken to the dorm, Jane Hall, where I was introduced to my roommate.

The next day was Sunday, and on Monday, I went through the registration process. Although I was a junior at the University of Liberia, most of my credits were not accepted, so I registered as a super-freshman.

CHAPTER 14

---◆◆ ◆ ◆◆---

First Experiences with Snow and Campus Life

My first weekend on MacMurray College campus was a lesson learned and a game-changer. My roommate, Kurt, was a very nice Caucasian and an excellent person. My first major experience with snow skiing was with Kurt. A few days after my arrival, Kurt suggested we go snow skiing on the football field about a mile from campus. Without fully understanding what it would entail, I got dressed and went with Kurt and his friends.

After about forty-five minutes of skiing on the field, my legs started to feel heavy. I told Kurt that I wanted to return, but he kept telling me to wait a little longer. After another thirty minutes, my legs were so heavy that I couldn't lift them. I dragged my feet back to the dormitory and made the mistake of opening the hot water faucet, inflicting major pain on myself and affecting my toes.

Kurt and I got along well, but there was one major issue: the females. They were always coming in and out of the room. The room was small with a bunk bed—Kurt had the upper bed, and I was on the lower portion. It became an issue, and I managed it until the end of the semester. At the beginning of the following semester, I decided to have my own room. Not wanting to let Dr. Goulding know the true story, I used other excuses, like Kurt getting angry if I turned the light on to study early in the morning. After much convincing, Dr. Goulding requested a single room for me in Norris Hall, where I stayed until my graduation.

There were a few friends who made my life easier on the MacMurray campus. Ricky was an African American who was quiet and down-to-earth. We spent many hours together discussing racism and its effects on African Americans. He was a philosophy major planning to go to divinity school after graduation. Another close friend was Jay, a Caucasian who lived in Norris Hall. His father was a preacher, and Jay was very nice. He often took me to his parents' place for weekends and short breaks. Jay was a linebacker on the football team and was also in the military. We spent many hours together, as both of us were majoring in Biology. Jay was always willing to defend me if someone tried to be mean to me.

The love people had for me on campus was amazing. I served as a teaching assistant in the Organic Chemistry Lab for two semesters. I participated in student politics, and became a Senator from Norris Hall, and then ran for President of the Student Government, losing by only three votes. That was a clear indication of my popularity on campus. I sang with the college choir, which took long trips during spring breaks to perform in various states. I also sang with the Grace United Methodist Church choir and was an active member of the Friendly Mixers, an older Sunday School group at the church. The youngest member might have been in their fifties, making me their "Little Angel."

Many families requested that I spend time with them during breaks. The two major families were Robert and Joan Leach and Walt and Dorothy Matthews. After Rob lost his wife in a vehicle accident, the Matthews stepped in and hosted me for weekends and short breaks. The Gouldings always ensured my well-being and safety were secured.

MacMurray College seemed destined for me because of the factors that made my life comfortable. The many families that God put in my path through the church and community were a blessing. One thing that always captivated me about this historic town of Jacksonville, IL, was the street named Liberia Street. I would jog once or twice a week and ride my bicycle every Saturday morning, covering over fifteen to twenty miles. One day, while jogging, I came across Liberia Street. I stopped and pondered why this street was named after my country. Despite asking around, I got no definitive response during my years in Jacksonville. Years later, in 2016, I took my fiancée and other friends to the site after MacMurray's 149th Commencement, where I received an Honorary Doctor of Public Service Degree. My fiancée took a picture of me pointing at the street sign.

Dr. Francis N. Kateh showing his family a street sign in Jacksonville, IL.

The two Honorary Doctorates Degree Awardees, MacMurray College, May 2016

With all the support from families, none compared to the love and support showered upon me by the Doyles. Many believe that God has an angel assigned to everyone, and there is an appropriate time to find yours. I think I had the perfect timing to meet my "God-sent" angels. Their positive actions toward me were incomprehensible, clearly the will of God.

CHAPTER 15

My Adopted Guardians

The Late John & Wilma Doyle, my God's earthly angels

John and Wilma Doyle were my guardian angels. One summer, John and Wilma, whom I affectionately called Dad and Mom, called the college to offer their home to a student who needed a place to stay during the summer. Shortly after that, during Easter break, John called again and learned that an African student needed a place to stay. They drove 142 miles from Henry, Illinois, to Jacksonville, IL, to pick me up. They were compassionate and treated me as their own child. John was a farmer, and Wilma was a retired teacher. They had a large farm with corn and soybeans planted interchangeably and also raised cows and pigs. Wilma was an excellent cook, making Henry my second home in the United States.

Our relationship went beyond what was anticipated. They didn't just provide a place to stay; they became my family and played a pivotal role in my journey to becoming a doctor. Even after I became a doctor and returned to Liberia, their parental care helped me make a significant impact at Ganta Hospital. They were truly God-sent.

Life at MacMurray College was good, and the campus atmosphere was very beneficial. Everyone worked together as a continuous and congenial community. The faculty was concerned if a student was not performing, and colleagues were willing to go the extra mile to ensure everyone understood the lessons and succeeded together. It was a community of good people that elevated each other. Every student was not just a number but a person who would one day make the college proud. This sense of community made life on campus accommodating and, most importantly, conducive to learning. Everyone became a family.

As a token of appreciation to the college, I did a "Bikeathon," riding my bike from MacMurray Campus to the Illinois Legislative Building in Springfield, IL, and back to Jacksonville, covering over 64 miles. I raised over $3,000, which was donated to the college to help construct an international plaque where the names of all international students would be published.

Mom and Pap Freiburg were the "parents" to all the students in the Biology Department. They were concerned about every individual and wanted each to succeed. They spent extra time with students who didn't perform well on exams, giving them additional work to ensure they understood the subject matter. All these factors made my life easier and my adjustment to a new country, different climate, and culture possible.

CHAPTER 16

---·◆◆◆◆◆◆·---

The Liberian Civil War

I adjusted well to life in the United States, but I never forgot my country. It was the day after Christmas when I was sitting with my guardians, the Matthews, in their living room. We were watching CNN when the news broke about an invasion in my native land by Charles Taylor. One international author described Charles Taylor "as a former civil servant in the Doe government [who] had been charged with embezzlement, fled to the United States, was jailed there, and escaped from custody." The author went on to write that "With an initial military force numbering around one hundred and with support from Libya's Muammar Qaddafi, Taylor attacked Liberia from neighboring Cote d'Ivoire on Christmas Eve 1989 and was able quickly and effectively to exploit the widespread hatred of Doe's government."[4]

At first, the civil crisis didn't bother me much because I believed that the United States would never permit the war to escalate beyond the point of entry. I religiously watched CNN every evening, expecting to hear that the crisis had been resolved. Unfortunately, I received the shock

---

[4] Soderlund Walter C. 2008. Humanitarian Crises and Intervention: Reassessing the Impact of Mass Media. Sterling VA: Kumarian Press, p. 2

of my life when a reporter interviewed Ambassador Chester Crocker, who was responsible for Africa at the State Department. He said that Liberia was not a critical partner to the United States and that U.S. citizens would be evacuated while Liberians should take care of themselves. I immediately began to shed tears. My friend Jay asked me what was going on. I lamented, saying that all my life, I thought Liberia was part of the fifty states of the United States and believed that nothing would ever happen to Liberia without her "mother," the United States, stepping in to remedy the situation. Hearing a State Department authority say otherwise broke my heart.

From that point on, my thoughts were always about how I could help my country. The war persisted, and I tried to contact as many relief agencies as possible to locate my parents and other family members. I was never successful until I graduated from MacMurray College.

CHAPTER 17

---◆◆◆◆◆---

Serving as a Camp Counselor

My first summer break at MacMurray College was uplifting. Based on my interactions with my colleagues on campus, my ability to work with younger kids, and my desire to ensure that I wasn't the only one to benefit from the scholarship, I spoke with Dr. Goulding about visiting some of my supporting churches to say thanks and to request additional funds to help other Liberians from the Liberia Annual Conference after my graduation. Dr. Goulding was excited, and this information was shared among the churches of the then Central Illinois Conference and now Illinois Great Rivers Conference.

Within weeks of my arrival, I was invited to the Annual Conference, where Bishop Woodie White gave me the podium to speak to the delegates. I expressed my appreciation and encouraged the churches to continue raising funds to support at least two Liberians at MacMurray at any given time. At the end of my speech, I sang "His Eye is on the Sparrow," which moved the delegates and made me a household name within the Central Illinois Conference. I raised enough funds to bring three additional students to MacMurray College after my graduation. Unfortunately, only one returned to Liberia, while the other two refused, leading to the closure of the scholarship scheme years later.

Recognizing my gift for working with various age groups, I was asked to serve as a camp counselor at East Bay Camp. While I felt I could make a difference, I was still concerned about how some of the kids would relate to me. To my ultimate surprise, I became a magnet for the younger ones. The camp director was astonished by the bond I formed with the children. I taught them African songs, introduced them to soccer, and took them canoeing on the beautiful lakes of Canada. At the end of that first summer, it became clear that my summers would be filled with more work. The following summer, enrollment at camp increased by over 15%, and most of the enrollees mentioned on their forms that they wanted to be at the site where I would be a counselor. It was a great experience for me and the kids, creating an indelible impression of non-discrimination and acceptance, demonstrating that in Christ, there is no east or west, north or south. We are all one, created in His image.

East Bay Camp, Illinois (June 1989)

Prior to my graduation, I took the Medical College Admission Test (MCAT), and my performance was acceptable. I applied to five medical schools, received calls from three, and had interviews with two: Meharry and the University of Southern Illinois. Both accepted me, but as a foreign student, I didn't qualify for any federal or other scholarships. My ambition to become a doctor hit a brick wall. It seemed I was only receiving scholarships to study divinity up to the doctorate level, but that was not my plan. My original American guardian, Dr. James Goulding, said to me, "Remember, your visa is going to expire, and MacMurray cannot protect your status. You need to accept one of these scholarships and become a preacher."

I responded, "Divinity is not my calling; I feel called to be a medical doctor. If that's the case, I have two options: go back to Liberia and be a farmer or keep trusting that my God will make the impossible possible so that I become a doctor." A few days later, I received a call from Betty Story, the editor-in-chief of the Current, the Central Illinois Conference of the United Methodist Church journal.

The late Mrs. Bettie Story and her lovely husband Rev. Dr. Story, Professor of Religion at Wesleyan University, Bloomington, IL

Betty said, "Someone told me about you, and I would like to come over and let your story be heard as to why you want to become a doctor." I was excited, and she came over and published an excellent piece in The Current. The article narrated my journey from the town of Karlokeh and my desire to become a doctor and return to Liberia to make a difference. As a student at MacMurray, I often served as a camp counselor during summer breaks and was loved by the campers. The seeds I had sown germinated, leading to calls from Methodist churches in the Central Illinois Conference wanting to hear my story. This yielded a testimony today. I had the opportunity to speak at thirty-one United Methodist churches and received commitments from twenty-three that came through to support me through an Advance Special managed by Dr. Goulding for my medical education.

When the commitments were made, and funds were being raised, the two medical schools willing to accept me had completed their admission process. Additionally, I was on an F-1 Visa, which disqualified me from student loans. My visa was set to expire a few months after my graduation, meaning I would have to forfeit my stay in the United States. Returning to Liberia was not an option due to the civil conflict ravaging the country. This was a major concern.

As usual, when it seemed like an obstacle was in my path, the Lord intervened. One afternoon, while preparing for my final exams, I got a call from Pap Freiburg. He said, "Guess what, Francis! I just researched and made some calls. You can still make it into medical school in the Caribbean. Come to my office; I'll give you the information."

"Wow! I'm on my way," I responded with excitement. I quickly walked from Norris Hall to the Biology Department and met with Pap Freiburg. He gave me the information, and I immediately began gathering the necessary requirements to apply.

I applied to Ross University School of Health Sciences in Dominica, the Commonwealth of the West Indies. The recruitment office was in Dearborn, Michigan, where I went for my interview and got accepted into the program. Upon graduation from MacMurray College, I left the United States for a new country I had never been to and knew no one from. With a leap of faith, I boarded the American Airlines flight to Orlando, Florida, and onward to Antigua, West Indies. From Antigua, I took a smaller flight to Roseau, Commonwealth of Dominica. Upon arrival, a bus awaited me and other students to take us to Portsmouth, Dominica, where the campus was located.

The ride from the airport was fearful due to the treacherous valleys with no guardrails to protect a vehicle from tumbling down hundreds of feet if there was a mistake. Additionally, they drove on the opposite side of the road, putting me in a panic. Finally, I arrived on campus and was escorted to my room, ready to embark on the next phase of my journey.

CHAPTER 18

———— ··◆◆·◆·◆◆·· ————

First Day in Medical School

We were fifty-seven students, and out of the total number, only seven of us were black. The first day in class was a frightening moment when the head of the Anatomy Department, Professor Twedele, walked in. He looked stern and authoritative. "Can all the black students please stand?" he said. Fear gripped me, but we all rose. He continued, "Good morning. I am Professor Twedele, and I have been teaching this class for the past seven years. Not a single black student has passed my course, so when you are failing, don't be afraid; you are not the first nor the last." The room fell silent. He then welcomed all of us, gave an overview of the course expectations, and wished everyone a successful semester. That was the beginning of another life experience.

Portsmouth was a great community, ideally located with cruise ships docking there because it was a tourist destination. Their economy was based mainly on tourism, but there were also banana plantations. I found the Methodist Church and quickly became part of the choir. That move helped me make friends and adapt to life on the island, which wasn't too different from Liberian culture.

The academic journey was intense. By midterm, one of my cadaver mates, Mohammed, committed suicide. Mohammed was a Pakistani whose parents, both physicians in the United States, wanted him to become a doctor at all costs. The stress was overwhelming for everyone. One morning, as usual, the bus came to pick us up. Mohammed walked towards the bus but didn't get in. He seemed deranged, and none of us, future doctors, realized it. After some persuasion, he got on the bus, and we all went to campus. I saw him later, heading towards the Anatomy Lab near the Caribbean Sea. I asked if he needed anything because I was going to do some dissection. Mohammed said no and mentioned he was going to the beach. That was the last time I saw him. He took off his watch and clothes and jumped into the Caribbean Sea. His lifeless body was found twenty-four hours later. It was a traumatic experience, and I felt guilty for not doing more to help him. This incident made me double down on my studies, knowing that failing Anatomy would jeopardize my sponsorship.

Anatomy was a challenging and costly course, priced at $3,750 USD. My grades were slightly below the passing grade of seventy-five percent, so I was panicking. A week before the final exams, I approached Dr. Twedele for guidance. "Prof, could you guide me on areas I need to concentrate on for the final exam?" I asked.

Dr. Twedele inquired about my cumulative grades. "Seventy-two-point-five percent," I told him.

"Wow, you will have to work harder because I don't scale grades. Study everything we covered over the semester," he said.

I left his office disappointed but determined to pass the course and prove to Dr. Twedele that I meant business. I stayed up for over seventy-two hours, memorizing and cramming every test and quiz I could find. I spent nights in the cadaver lab alone, drinking over twenty-five cups of coffee to stay awake.

On the day of the test, I prayed and went forth. There were two hundred and fifty multiple-choice questions for the three-hour written exam, followed by a hundred practical exam questions on the cadaver for two hours. Everyone had to come up with a code under which their scores would be posted. My code name was "Thank you, God." After the exams, I had a negative reaction to the coffee; my lips, eyelids, and face swelled up. I was at home taking antihistamines when my close friend Woozevelt Pierre from Haiti came over to give me the bad news of all those who failed. Woozevelt didn't know my code name, so he wrote down all the failing codes and came over to me.

"What is your code?" he asked. I couldn't remember. Woozevelt then read out all the failing codes. When he finished, he said, "You can't remember?"

"No, I just can't," I replied.

Then Woozevelt said, "There was one person who scored the highest on both the written and practical exams, and the code is 'Thank you, God.'"

I jumped up and exclaimed, "That is me!" I had a cumulative score of ninety-four percent.

The next day, I went to campus and met with Dr. Twedele. He ushered me into his office and said, "Had someone else graded your paper, I would have challenged the grades, but I graded it myself. Thank you. You were the only black student to pass the course."

Feeling boastful, I said, "Remember, I am not just a black student; I am indeed an African-African."

Dr. Twedele responded, "Now I know."

Ross University was exciting, but after the first academic year, which was a trimester system, the school authorities decided to increase the tuition by almost fifty percent. With my dependence on others for survival, I realized I couldn't manage the increased costs. My friend Woozevelt Pierre told me about Spartan University Health Sciences in Vieux Fort, St. Lucia, where he had transferred after failing Anatomy at Ross. Spartan's tuition was about half of Ross's new rate. Given the limited funds coming in through the Advance Special, I quickly requested a transfer to Spartan Health Sciences, where I completed my basic sciences.

And so, another chapter in my journey began, full of challenges but driven by determination and faith.

CHAPTER 19

---••◆••---

Life in St. Lucia, West Indies

Moving to St. Lucia brought a completely different experience. I was fortunate to live in a very nice residential area that included a mini mart owned by Bertha Lewis. Bertha was incredibly kind and accepted me into her home not just as a tenant but as a son. At times, she even provided meals and was always concerned about my well-being. Bertha was well-connected in St. Lucia, a close friend of the Prime Minister, the late Hon. John Compton, who was a practical and down-to-earth person.

At Spartan Health Sciences University in Vieux Fort, St. Lucia, I adjusted well and became one of the highly respected students. I was very competitive and focused on achieving my dream of becoming a doctor. Unlike the stress of getting to and from campus in Portsmouth, Dominica, I had additional resources here. I bought a 275-horsepower motorcycle with a radiator, a very powerful bike. After using the bike for several months, I purchased my first vehicle, a black, two-door Mitsubishi Electra sedan. I became the talk of the campus.

I spent some Saturdays visiting the beautiful and historic sites in St. Lucia, like the great Pitons, the Sulfur Springs, or Castries, to see the various cruise ships arriving or departing. After spending a year plus in

St. Lucia and completing my Basic Science Courses, I had to continue my clinicals in the US, Mexico, or somewhere else. I was accepted to do my clinicals at Norwegian American Hospital in Chicago, Illinois.

Needing a visa to continue my education in the United States, I flew to Barbados, where the nearest US Embassy was located, after booking a visa interview. I went through the interview, and there was only one question posed to me. "When was the last time you visited Liberia?" I responded, saying it had been over five years. The consular then asked, "How am I sure you would return to Liberia?"

I immediately said, "There is nothing I can say to convince you, but my commitment to my country is a covenant between my mother, myself, and God. Nothing, not even the war, can stop me from returning home when I am done with my education." The consular immediately stamped my passport, "DENIED." It was devastating and traumatic for me. It seemed my entire world had crumbled. With tears running down my cheeks, I politely walked out, spent the night at a hotel, and then flew back to Castries, driving to Vieux Fort the next day.

Upon arrival in Vieux Fort, St. Lucia, my landlady Bertha asked if I got the visa. With tears in my eyes, I told her, "No."

"Go and rest. I will see you later," she said. I didn't know how I fell asleep. The next morning, there was a knock on my door. It was Bertha. When I woke up with tears still in my eyes, she was moved with sadness and said, "Don't worry; it will be well. I spoke with my friend, the Prime Minister, and he said he would need to see you and see how he can help you obtain the visa. All is not lost; you can't give up hope," she told me.

About 4 pm that same day, Bertha came over to my apartment and said, "The Prime Minister would like to see you very early; therefore, we will need to leave by 4:30 AM." I got up by 2:00 AM and got myself ready in a three-piece suit. I came downstairs by 4:00 AM and was just pacing around the front of the house when she also came down and saw me in my suit.

"Where are you going with that on?" she asked.

"Oh, you said we were going to see the Prime Minister, so I have to dress up."

She laughed hard and said, "Please go back and change to your short trousers and a T-shirt."

Wow, I thought, that is not appropriate to meet the Prime Minister. But I went back upstairs and changed. As usual, I loved to be casual. I wore short pants and a muscle-arm T-shirt, and off we went.

As we drove towards Castries, she instructed me to turn left on a dusty road. I became a bit confused. "Where are we going?" I asked quietly.

"Just continue," she said. After driving for over twenty minutes through a banana plantation, we came across a Nissan single-cabin pickup. We saw a man standing by the road, and she told me to stop. She asked the man, "Where is your boss?" He responded, "He is just ahead, cleaning among the banana trees as usual. You know him," the man said, speaking Patois, a broken French language created by the locals. They laughed, but I didn't comprehend what was unfolding before me.

We got down from the vehicle and walked towards a man in jeans and a shirt with a machete in his hand. She whistled to him, and he responded with a whistle and then said, "Wow, you brought the young man? Please give me five minutes; I will be with you both." We both sat on a log stool and after a few minutes, the Prime Minister came and greeted us.

"How can I help you, young man?" the Prime Minister asked after speaking Patois with Bertha.

"Which country in Africa are you from?" he asked.

"Liberia," I told him.

The Prime Minister bowed his head for a few moments and said, "I am so saddened about what is going on in Liberia. I was planning to visit Liberia during her Independence when I heard about the assassination of my friend and brother, Dr. William R. Tolbert. The country has lost a visionary and a patriot. I am not surprised by what is going on there now. But how can I help you?"

I explained my situation, "Sir, I just completed the required basic sciences in medicine, and I need to do my clinicals in the United States. I got accepted at Norwegian American Hospital in Chicago, Illinois, but was denied a visa because the consular felt I might not return to Liberia after graduation due to the crisis."

The Prime Minister listened intently. I continued, "Sir, my desire to be a doctor is because of my mother. She can't read or write because she lost one of her eyes at a young age. Her peers teased her, so she didn't go to school. I asked her at age twelve why she had one eye. She explained through tears how her eye was infected, and because there wasn't a western doctor in the village, an herbalist placed herbs in her eye, and she lost her sight. Sir, I honestly want to be a doctor to help others."

The Prime Minister moved with compassion and requested my full name, date of birth, and letter of acceptance at Norwegian American Hospital. I gave him the information and expressed my heartfelt appreciation. The Prime Minister said, "I am proud of you, and you will make a difference because of your love for your mother."

Bertha and I left for Vieux Fort, St. Lucia. The car was quiet, reflecting on the conversation with the Prime Minister. Bertha broke the silence, saying, "It was a great moment with the Prime Minister."

"Yes, indeed," I responded.

Bertha added, "I have known the Prime Minister since we were kids, but I have not seen him so emotional as when you were talking about why you want to be a doctor. His eyes were red as if he was about to tear up."

"Wow," I said, "I wasn't paying attention because I was dumbfounded to see a Prime Minister with a machete without dozens of security and armored vehicles protecting him. That is humanity at its highest level. I wish to emulate him one day."

We arrived in Vieux Fort after almost two hours and went to our respective domiciles. By 3:00 PM, Bertha sent her daughter to call me because there was a long-distance phone call for me. I ran downstairs to take the call. It was from the US Embassy in Barbados, requesting me to return for another interview in two days. "Thank you so much," I told Bertha. "You are indeed a rescue mother. I can never repay you for what you have done for me."

On that memorable Thursday morning, I drove to the airport and flew to Barbados. Upon arrival, I took a cab to the US Embassy. I was ushered in for another interview. This time, the atmosphere was more welcoming, and there was only one question. "What is your tribe?"

"I am Kru," I answered.

The consular followed up with a strange question, "What is the relationship between the Kru and Krahn?"

"Wow, I really don't know," I explained. "When the coup occurred in 1980, there was a Kru Colonel called Jebboh who it was claimed would have headed the interim government if successful in staging the coup in Bentol. Unfortunately, President Tolbert spent the night at the Executive Mansion and was assassinated. Colonel Jebboh was killed crossing the Cavalla River a couple of months later after he went into hiding because the head of the interim government considered him wanted."

The consular said, "Thanks, you are knowledgeable." He then asked, "How long do you want to stay in the US?"

"Two years," I said.

He issued me a three-year visa. I appreciated the consular and asked, "May I ask you a question?"

"Please, go ahead," he said.

"Did you ever live in Liberia?"

He smiled and said, "Yes, I lived in Liberia for over seven years. I was the head of the Consulate Section there."

"Thanks again," I said, and as I walked out, I was grateful to God for being faithful. This time the night felt incredibly short because I couldn't wait to board the flight to St. Lucia. I was eager to express my thanks and appreciation to Bertha, whose intervention had made it possible for me to have another opportunity with a "staged interview."

The next morning, I boarded the flight to Castries, St. Lucia. Upon arrival in Vieux Fort, I wasted no time and rushed to find Bertha Lewis. When I saw her, I immediately began expressing my heartfelt gratitude. I thanked her profusely for her pivotal role in creating this opportunity for me. I also asked her to convey my deepest thanks and appreciation to the Prime Minister, Hon. John Compton, for his direct intervention in making this possible.

CHAPTER 20

———— ·◆◆◆◆·· ————

Back in Chicago, Illinois

It seems Illinois was destined to be the place where my dream would come true. My first scholarship came through the Illinois Central Conference, now the Illinois Great Rivers Conference. Chicago has a rich history intertwined with numerous positive initiatives for those considered at the lower rungs of life. For me, it felt like my life's journey in achieving my ambition was woven into the fabric of Illinois, where I was positively embraced. Returning to Chicago was a continuation of what began over five years earlier when I first touched down at O'Hare International Airport.

This time, my arrival in Chicago was unremarkable. Upon arrival, I stayed with my friend Woozevelt Pierre for a month while I searched for my own place near the hospital. Norwegian American Hospital, situated near Humboldt Park, became my new academic and professional home. At Norwegian American Hospital, I encountered great mentors like Drs. DeLeon, Gonzales, Munoz, Park, and Burriosou. They had a profound impact on me by providing both theoretical and practical aspects of medicine.

Dr. DeLeon stood out among all the doctors at the hospital because my focus was always surgery. As a surgeon, Dr. DeLeon pushed me to the limit to understand the importance of surgery. We spent countless hours together in the lecture room, operating theater, and clinic. Surgery was always fun for me. Because of the bond we built, Dr. DeLeon recommended me to his sister, Dr. Parks, who had her private facility on the East Side of Chicago, where I moonlighted to make some extra money.

Dr. Parks, her Administrative Assistant and I, 1994

While most of my clinicals were done at Norwegian American Hospital, there were a few courses I completed at other facilities, such as St. Anthony Hospital, where I did Pathology, and Cook County Hospital, where I did Emergency Medicine as an elective. The elective at Cook County Hospital sharpened my skills in trauma and taught me the efficiency needed in dealing with emergency cases. This extra knowledge would be crucial when I returned to Liberia, especially in the war zone.

Returning to Chicago felt like coming full circle. It was here that my journey towards becoming a doctor had begun, and it was here that I would refine my skills and gain the experience needed to make a difference. Illinois, with its history of embracing and uplifting those who strive for a better future, has become a significant part of my story. My mentors, friends, and the opportunities I found in Chicago helped shape me into the doctor I had always dreamed of becoming.

CHAPTER 21

---◆◆◆◆◆---

The Call I Knew Would Come One Day

I had less than six months left in 1995 to complete my studies. After a seventy-two-hour assignment at the hospital, I came home hoping to rest, but I was restless and couldn't fall asleep. Just when I thought I might finally drift off, the phone rang. It was an operator asking if I would accept a collect call from Ivory Coast. I said, "Yes."

The caller was my younger brother, with whom I hadn't spoken in over five years. The first question I asked was, "How is Mom?" My brother became very quiet, and instantly, the thought of my mother being dead entered my mind. I asked, "When did she die?" and burst into tears. This was something I had often felt—that I would not return to Liberia to see my mother alive. The first time the thought struck me was when I was boarding my flight from Liberia to the United States. As I walked towards the aircraft, I felt I would not see my mom again, and now the reality was at hand.

After crying, I told my brother that I would send them some funds and would plan a visit to the Ivory Coast to see the rest of the surviving family. As promised, I arranged a visit and returned to Africa to bring some closure to this painful part of my life. Meeting my dad, brother, and other relatives and friends was a great reunion, but my mother's absence haunted me daily. I was always quiet and often pondered what my mother would have said to me since I was only a few months away from achieving our shared dream, inspired by her affliction.

After a week in Ivory Coast, I made a visit to Monrovia. My late friend Fred Logan made the trip possible. Fred was the next student to attend MacMurray College after my graduation, based on the funding I had raised. Upon his graduation, he returned to Liberia and worked for the United Methodist Church as the Treasurer. I was grateful to Fred for making it possible for me to rekindle bonds with friends who had survived the war. Returning to Liberia brought a mix of despair and rededication. The ambivalence I felt was resolved: I knew I would return to Liberia after graduation. My country needed me.

Seeing the country from a different perspective was a wake-up call. When I left Liberia, it was calm and peaceful, although there were issues of dissatisfaction among various political stakeholders and students, especially the Vanguard Student Unification Party (SUP), which pressured for the rule of law and the upholding of everyone's rights. Driving around Monrovia, I saw the devastation and remnants of the war, with buildings infested with bullet marks. I also saw frustration, depression, and hopelessness on the faces of many friends and colleagues. These sights weighed heavily on my mind as I went to bed each night.

This was a wake-up call. My profession as a doctor was to be a beacon of hope, and it was prudent that I return to help in any way possible. My love for my country was never extinguished, only intensified by what I saw post-war. The peace and harmony I remembered were now replaced with new challenges. I realized how blessed I was to have left the country before the war, and I felt a deep sense of gratitude for being alive, unlike those who lost their lives. The best thing I could do was to return to the United States, complete my studies, and then come back to contribute to the rebuilding of Liberia by providing healthcare services.

CHAPTER 22

---◆◆◆◆◆·---

Returning to Liberia

About a month before my graduation, I faced my first real temptation. I received an offer to work with my mentor, Dr. DeLeon. It was a difficult decision, but I thanked him for his mentorship and explained why I went to medical school. I reiterated the commitment I made to my mother and God and how I had to return to Liberia to provide care to those in need, especially in the wake of the humanitarian crisis that claimed over 250,000 lives. Dr. DeLeon understood and even helped me ship some personal belongings to Liberia.

Some classmates also tried to convince me to stay and open a practice with them, but I declined. The United Methodist Church of Liberia purchased my ticket, and I arranged for my family members in exile in Ivory Coast to meet me in Abidjan so we could travel to Liberia together and reunite after years of separation.

I bought four tickets for my father, brother, and two nephews, assuming they would meet me at the airport. When I arrived and waited for over three hours with no sign of them, I left for the hotel, feeling worried. As I was about to order dinner, a man approached me and asked if I was the doctor from the United States heading to Liberia. He said he saw my brother and cousin at the airport. I immediately took a taxi back to the airport and found them.

"Where is the rest of the family?" I asked.

"Big brother, there's a long story," my brother said. "Please, let's get to the hotel, and I'll explain."

At the hotel, my brother explained that they were indebted to some people, and the funds I sent weren't enough to cover both the bills and transportation costs. I felt a wave of frustration.

"Wow! This is unfortunate," I said. "We talked daily, and neither you nor our father informed me of your financial situation. Now, we might forfeit the tickets I purchased because they are non-refundable. We need to try and see if we can change the date instead of forfeiting all the tickets."

My younger brother then dropped another bombshell. "No big brother, it's more than that. We are not four but eight."

I was taken aback. "Wait! What?" I decided to stay calm. "Let's eat, and we can talk early tomorrow."

That night, I barely slept, trying to figure out how to extend the tickets and what the additional cost would be. The next morning, we went to the airport, but it was difficult to convince the agents to assist in deferring the tickets. Frustrated, I gave my brother enough money to cover the debts and any other contingencies. I left for Liberia, hoping to send more money to transport them to Monrovia by vehicle.

Upon arriving in Liberia, I went to the United Methodist Church headquarters to begin the process of legitimizing my profession. Bishop Rev. Dr. Arthur F. Kulah was out of the country, so I met with the Administrative Assistant to the Bishop, Rev. John Innis. He welcomed me warmly and appreciated my decision to return and support the church. I explained my commitment to humanity, the church, my mother, and God. Rev. Innis prayed for me and prepared a letter for the Ministry of Health along with my documents.

I was required to do a year of internship at the John F. Kennedy Memorial Medical Center since I had studied abroad. I respected this decision and began my internship.

About a month into my internship, I was urgently called to the United Methodist Church Center by the Bishop. I took a taxi to the bishop's office and was ushered in. The bishop left his desk, placed his hands on my shoulders, and said, "Take courage."

I was confused and asked, "What did you say, Bishop?"

He explained, "I just got the news that your father, brother, and nephews drowned around Sinoe County while en route to Monrovia from Tabou, Cote d'Ivoire."

Stunned, I asked, "How can I get to Greenville, Sinoe?"

The Bishop said he would negotiate with the Peacekeepers to take me there on their flight. After three days of waiting with no progress, I decided to go to the ECOMOG Peacekeepers base.

As I arrived, a relative told me, "Your father is in Duala, at your cousin's house."

I was astonished. "I was told he was dead; how is that possible?" I took another taxi to Duala and found my father with my two nephews and a niece. I had lost my brother, two nephews, and my brother's fiancée. What a pity.

After making all the necessary arrangements, I resumed my internship, reunited my family, and took on the role of the breadwinner. Despite the immense challenges, I was determined to fulfill my commitment to my family, my country, and my faith.

CHAPTER 23

————— ◆◆◆ —————

What Happened?

As I was told, upon the return of my brother and cousin to Tabou, Cote d'Ivoire, my dad decided to use the funds to pay off their debts. They didn't want to put an extra financial burden on me, so they opted to travel by canoe operated by the Fanti seafarers. Unfortunately, the canoe was overloaded with both passengers and goods. Despite reservations about the decision, they left the shores of Tabou. As they approached Greenville, the Liberia Peace Council (LPC), led by George Boley, chased the canoe, forcing the conductors to go further out to deep sea. A heavy wave caused the canoe to capsize. Out of the eight family members onboard—Dad, Anthony, Mark, Bartequa, Jugbeh, Jugbeh's mother, and two brothers—only four survived. My brother, his fiancée, and their two kids drowned.

In the months that followed, my passion for medicine grew stronger. I was determined to apply the theoretical knowledge I had gained to help bring relief to many. One afternoon, while covering the Emergency Room with other physicians, a patient was brought in with a ruptured ectopic pregnancy. The two physicians who initially saw her seemed to use the case to entrap me, possibly out of envy since I had come from the United States. After completing an assessment on another patient, I heard this patient crying. When I asked where the other physicians were, the nurse aides responded in unison that they didn't know. I immediately went to assess the patient, but her history and information were vague because she was going into shock, and the caretaker couldn't provide substantial details. I ordered labs, and the hemoglobin result came back at 6.4 gm. Something was seriously wrong.

Meanwhile, the two doctors had informed the consultant, the head of the John F. Kennedy Memorial Medical Center, about the suspected ruptured ectopic case and claimed they had another emergency outside the hospital.

The consultant asked, "So, who is covering the emergency room?"

They replied, "Oh, the American doctor."

While trying to make sense of the patient's condition, one of the nurse aides suggested, "Doc, can you take a needle and place it in the patient's abdomen to see if there's blood?" The patient's abdomen was distended as if she were twenty to twenty-four weeks pregnant. Realizing the nurse was right, I combined theory and practical knowledge, performed an abdominal tap, and aspirated blood that didn't coagulate due to mixing with intra-abdominal fluid. The inability of the blood to coagulate indicated active bleeding. Given the patient's childbearing age, additional inquiry was required. Since the patient was non-communicative, an interpreter was brought in to ask the relative about her menstrual history. The relative revealed the patient had not seen her menses for over two months, confirming a suspected ruptured ectopic pregnancy.

I called the consultant using the walkie-talkie, informing him of my findings. He bombarded me with questions, but while answering, the nurse aide and I had already put the patient on a stretcher and were rolling her to the operating room. At the Maternity Center, which had the only functional operating rooms, I prepped the patient, a task I was accustomed to from my training in the U.S. When the consultant arrived, the OR nurse informed him that the patient was prepped. He entered, surprised, and asked who taught me to prep a patient. I told him it was a routine in the U.S. He nodded, impressed. We performed a salpingectomy (surgical removal of the fallopian tube), maintained hemostasis (stopping the flow of blood), and did a peritoneal lavage (cleaning the abdomen with saline mixed with antibiotics). The consultant then asked if I could close using an uninterrupted stitch on the peritoneum and a figure-eight stitch on the fascia. I confirmed, and he observed for a few seconds before saying, "Continue, doc; you're doing a great job."

What was intended to undermine me turned into a victory. From then on, the consultant had more confidence in me and occasionally sent me to his private clinic to see patients. Despite the envy and trials from other physicians, God saw me through victoriously. My willingness to listen to everyone—from the cleaners to the patients—and systematically address their health needs while empathizing with them gradually made my name a household one.

CHAPTER 24

————·◆◆ ◆ ◆◆·————

The Infamous April 6, 1996 Crisis

After seven years of a brutal civil war, Liberia's political, religious, and military leaders reached an agreement to form a Council of State, a "collective presidency" where warlords and other key figures shared power. A noncombatant civilian chaired the council, while warlords and civilians served as members. This fragile peace shattered when Charles Taylor, one of the council's "presidents," decided to arrest Roosevelt Johnson, another warlord. The city exploded into a full-scale urban war.

The war saw bizarre alliances, with former enemies joining forces to fight a common foe. ULIMO K, NPFL, and even the Ghanaian contingent of Peacekeepers initiated the conflict. It all started at 1:00 AM on April 6, 1996, while Dr. Emmanuel Sandoe and I were on call at the Emergency Room at the John F. Kennedy Memorial Center (JFK). A patient's relative, who had stepped outside for fresh air, rushed back to inform us about Ghanaian troopers taking up positions around the medical center. We went out and indeed saw soldiers positioning themselves. We asked to see the commander, who assured us they were conducting a simple mission and that we, along with the patients and staff, would be protected. His reassurance calmed us, and we informed everyone inside that there might be some shooting, but we should be safe.

At exactly 3:00 AM, a loud sound echoed—the launch of a Rocket Propelled Grenade (RPG). Continuous gunfire followed, directed towards the Atlantic Ocean side of Sinkor 19th and 20th Streets. The bombardment persisted until 6:00 AM when Roosevelt Johnson's forces counterattacked, with an RPG landing near the JFK compound and setting a building on fire. The exchange of fire intensified as the combined troops of ULIMO K, NPFL, and the Ghanaian Peacekeepers retreated. ULIMO J, led by the infamous "General Butt-naked," left their compound, crossing Tubman Boulevard towards the Barclay Training Center (BTC). It was a surreal and chaotic scene, with fighters, some with minimal military hardware, moving majestically while chanting.

The days that followed were chaotic, with mass casualties, including elite special forces of the National Patriotic Front like General Wheyee. What puzzled me most was witnessing cannibalism. I saw the thorax of a ULIMO J member, killed and brought to the old YES Gas Station at the Old Road junction, being opened by a female fighter. She removed the heart, shredded it into pieces, and ate it, along with others. I was in shock, unable to comprehend what I had just seen—my countrymen and women eating another person's heart.

As I stood there, a general recognized us as physicians and called us over. He assured us that it was safe to return to the hospital. After a long delay, we received the all-clear to go back to JFK. Upon arrival, we found that some patients had fled due to the fighting, while others who were critical remained. Dr. Sandoe and I quickly triaged and began providing care.

During the evening hours of April 7, 1996, General Benjamin Yeaten, commonly called "50," arrived at the hospital. He wanted to know who the doctors were. Guided to the ER, he met Dr. Sandoe and me. General Yeaten told us that "President Taylor" had sent him to ensure the hospital remained operational. He asked who was in charge, and Dr. Sandoe pointed to me. I was nervous but thanked him and outlined what was needed: fuel for the generators, food supplies, and incentives for the staff who hadn't been paid for over five months. I also stressed the need for a direct link and protection for the safety of patients and staff. General

Yeaten left with his team and returned within two hours with 1,500 gallons of diesel, 100 bags of rice, and assorted supplies. It was a great start.

Both fighters and hospital staff played their roles well. Services were restored under tense conditions, and we faced constant looting and harassment. The fighting became more intense, with numerous casualties.

One of the most intense moments occurred when five wounded fighters were brought into the ER. General Tarnue appeared and began interrogating them about their injuries. The fighters were terrified and refused to speak. Tarnue pulled out his pistol and threatened to shoot them. I stepped between him and the patients, pleading for mercy. Trembling, the general agreed to leave them alone but insisted he never draws his pistol without using it. Three fighters fled, and I don't know if they survived. The other two had life-threatening injuries.

We later learned the fighters had been looting near Charles Taylor's house and were mistaken for enemy combatants by their colleagues, who then opened fire on them. They evacuated them under the pretense of coming from the battlefront. Despite the dangers and risks, we, the health workers, continued to provide services to both civilians and fighters caught in the battle.

My ER rotation at Cook County Hospital in Chicago had prepared me well for this, giving me the skills needed to save lives. Dr. Peter Coleman also played a pivotal role in developing our surgical skills, making us confident in performing life-saving procedures. Despite the unfortunate situation, it enabled the young physicians at the hospital to gain valuable experience and build confidence in their abilities.

CHAPTER 25

———— ·◆◆◆◆◆·· ————

The Ultimate Risk to My Life

On April 29, 1996, Charles Taylor decided to make his move to the Executive Mansion. How ULIMO-J got wind of this remains a mystery, but they successfully prevented him and his entourage from reaching their destination. The exchange of gunfire and artillery could be heard as if it were happening right outside, around 16th and 17th Street Sinkor, just a block or two from the hospital. Dr. Coleman and I were in the middle of an abdominal surgery on a woman who had been hit by a stray bullet. Her abdomen was opened from just above the pubic area to a few centimeters below the sternum. As we were completing the ileostomy, another stray bullet came through the operating room, hitting the wall.

Fearing for our lives, Dr. Coleman, the rest of the medical team, and I had to leave the patient on the operating table. The fighting had intensified, with loud artillery fire approaching the main hospital at the Japanese Maternity Center. Those medical staff who lived off the JFK Compound, including Dr. Coleman, quickly escaped through the back entrance near the intern dormitory. I returned to the intern dorm where other physicians like Drs. Passawe, Sandoe, Dada, Gayflor, Kamara, Sieh, and Beh were staying.

Earlier that morning, my fiancée had a dream and woke me up to pray. She dreamt that I was captured as an enemy and was about to be killed when suddenly a man appeared, insisting that I wasn't an enemy and needed to be released. After a prolonged argument, I was set free. She warned me to be very mindful at work.

When the incident occurred at the hospital, and I returned to the dorm, my fiancée thanked God. I was concerned about the patient we had left on the operating table and others who couldn't leave due to their conditions, so I decided to go back and check on them. I first went through the emergency room and realized most of the patients had fled. Dressed in my scrubs, I cautiously walked towards the Japanese Maternity Hospital, which was now being used as the main hospital.

As I approached, I heard a voice command, "Halt! Who are you?" I responded, "I am a physician, and I'm trying to check on my patients." A man with an AK-47 rifle advanced toward me. Standing under a mango tree, not knowing what to do, I silently prayed. It was after 7 PM and partially dark, so I couldn't see who it was. When he got closer, I showed him my identification card and introduced myself. He jerked my flashlight and said, "Oh, you're the American Doctor. I need your heart."

I immediately froze. Just two weeks earlier, I had seen someone's heart ripped out and eaten, and it seemed like a possible replay. The unidentified fighter pulled out a knife from its sheath attached to his right foot and attempted to stab me. "I'm just trying to check on the patients left at the hospital, especially the one we were operating on!" I shouted. The fighter and I tussled over the knife for what seemed like an eternity. Just when I was getting weak, I heard a voice say, "I'm holding his hands. Run!" I ran into the hospital and hid in an opening near the panel box on the left side of the walkway.

I stayed hidden for over an hour. When there was no noise, I quietly turned on my penlight to find my way out since it was dark. To my amazement, a mentally impaired woman was also in the area where I had been hiding, and she hadn't said a word. I got out and returned to the dorm. I explained the episode to my fiancée, who immediately praised God for the intervention. One of the staff who heard and witnessed part of the episode was a young physician assistant who now practices as an ophthalmic nurse at J. J. Dossen Hospital in Maryland County, Liberia.

The next morning, I called my guardians, John and Wilma Doyle, in Henry, Illinois, and told them the story. They immediately wanted me to leave Liberia. "I am going to get you a ticket to leave Liberia. It is not safe, and you can't risk your life," Dad said.

"Ok, Dad," I replied, "But please wait and let me call you tomorrow to see if you can purchase the ticket." It took about a week before I could call back due to the difficulty of accessing a telephone booth. When I finally got through, I told them, "Dad and Mom, it's ok. God is in control. I don't intend to leave. If God wanted me to die, I would have died. I am sure He has a purpose for me to fulfill. I am willing to listen to His guidance in fulfilling what He wants me to do."

"Ok," my guardian said. "We got you and will continue to pray for you. Please be safe and let us know if there's anything you need from us."

After that incident, the Peacekeepers brought the situation under control, and an interim leadership was put in place that would lead to elections. I completed my internship at the John F. Kennedy Memorial Center, where I became closer to Dr. Coleman and other senior colleagues. My dad sent me some funds, and I purchased a white Daihatsu Charmant with a long antenna on the back trunk. The license plate on it read DUKA-1.

When I received the funds, I decided to buy a vehicle to enable me to go to Ganta when I wasn't on schedule at JFK, to provide services to the people of Nimba and its environs at the Ganta United Methodist Hospital. I wanted a specific but captivating name that would draw attention. During the process of elimination, I thought about my birth month, July, and its symbol, the crab. Growing up, I lived near a swamp where the most delicious crabs resided. The Kru word for the hairy crab is "Duka," hence the name. To make my fiancée happy, as she was a bit agitated about the name, I creatively told her it combined part of her surname (Dunn) and part of mine (Kateh), forming the name DUKA, the most amazing, hairy, and delicious crab.

CHAPTER 26

———•◆◆ ◆ ◆◆•———

General DUKA

With my new vehicle, I began traveling to Ganta Methodist Hospital every time I wasn't on call. Going to Ganta in 1996 was quite challenging. There were more than fifty checkpoints on the road, making a three-hour drive take close to ten hours. During my third visit to Ganta, I met General Stanley, one of the most feared and formidable generals. He was small in stature, fairly light-skinned, quiet but very calculative. On this visit, Patrick Mantor and Harry Wonyene, two staff members of the Methodist Mission, took me to General Stanley's Club. As I sat there, observing and drinking a Malta, General Stanley walked up to me and said, "You look out of place here."

Mr. Harry Wonyene and Dr. Francis N. Kateh (Ding, Ding and Dong Dong) Administrator and Chief Medical Officer, Ganta United Methodist Hospital (1998)

Mr. Harry Wonyene quickly approached and said, "General, this is the new doctor I told you about. He will join us soon."

"Wow!" the general exclaimed. "He looks very young and seems quiet. You mean he doesn't drink alcohol?"

"No, he doesn't," Mr. Wonyene replied.

"That's very interesting," the General said with a smile, shaking his head before returning to his strategically positioned seat to observe everything and everyone. After about thirty minutes, one of his bodyguards approached and said the General wanted to see me. Mr. Wonyene was very concerned but reassured me that it was okay; the General had found favor in me and only wanted to talk.

General Stanley took me to a private area and said, "Doc, I don't know why, but I like you and want you to be my friend." In return, I expressed my appreciation and accepted his friendship. We discussed some classified information, and at the end of our meeting, he asked, "Did you have trouble getting through all the checkpoints on the highway?"

"Yes, General," I replied.

"From now on, you should have no more problems. Do you have your own vehicle?"

"Yes, General," I responded.

"Good. When you're driving, keep your windows up with your blinkers on, and as you approach a gate, flash your high beams twice. That will signal that you are a two-star general. The gate will be dropped immediately."

I thanked him, and he escorted me back to my seat in the club. Returning to Monrovia that Sunday was surprisingly easy. It took less than four hours from Ganta to Monrovia. I had formed a long-lasting relationship with General Stanley, which lasted until I returned to the United States after five years of friendship.

Due to the unique name, DUKA, and my frequent presence on the highway, most of the security personnel at the various checkpoints later became part of the rebranded security sector after Charles Taylor overwhelmingly won the election in 1997. When I bought a new car, a red Toyota Celica, I transferred the plate, making me easily recognizable as a two-star General.

CHAPTER 27

———— ◆◆◆◆◆· ————

Moving to Ganta: A Dream Becomes a Challenging Reality

After completing my internship at JFK Medical Center in Monrovia, I requested an assignment to Ganta Hospital. This hospital had played a pivotal role in shaping my dream of providing quality healthcare to those in need. On January 18, 1997, I made the final leap toward achieving my goal of giving back to Ganta Hospital, which had accepted me as a volunteer and where Dr. Walter Stephenson and Lois Kohler, both missionaries, had intervened to change my destiny. Additionally, it was through Ganta Hospital that Bill Warnock, another missionary, made the final push for me by purchasing my ticket to the United States when the local church I served couldn't help. I was determined to reciprocate the kindness shown to me by making a difference in the lives of my people. Filled with gratitude, I was finally returning to Ganta, where my dream of becoming a medical doctor began to take shape.

Upon my arrival at Ganta Mission as a physician, I received my first shock. I arrived around 8:00 PM after a long drive from Monrovia in my red Toyota Celica Coupe with the license plate DUKA 1. I was escorted to my residence, and to my dismay, the entire house was empty—no mattress, bed, table, chairs, or utensils. Standing in the middle of the two-bedroom apartment, I pondered how I would spend the night. After a while, I spread a few pairs of jeans on the floor and slept.

The next morning, I spoke with Mr. Harry Wonyene, who took me to a Lebanese businessman named Bashir Hammoud. Bashir trusted me and opened a line of credit, allowing me to obtain a mattress and a set of pillows. Instead of feeling angry or demoralized, I used every unfortunate situation to build relationships. More than 25 years later, Mr. Hammoud and I remain friends; our bond has grown into something positive. With this initial experience, I quickly realized that working at the hospital would come with its own set of challenges, especially with my boss.

Mr. Bashir Hammoud, the Business man that trusted me without knowing me in making my stay in Ganta a bit comfortable

Dr. Barcolleh was a Russian-trained physician who worked hard and was the only physician at the hospital immediately after the civil crisis. I soon realized that my arrival would be challenging for him. The hospital had been frequently looted, and patient beds were in short supply. Relatives often had to bring mattresses for their patients. Surgical instruments were limited, requiring innovation to achieve successful outcomes. Despite these challenges, Dr. Barcolleh worked tirelessly to provide the best care possible with the available resources. However, my arrival seemed to unsettle him, as he likely viewed me as a threat to his authority.

On Monday, January 20, 1997, I showed up at his office with my letter of assignment. Dr. Barcolleh welcomed me but expressed uncertainty about whether I would accept the offered salary due to the hospital's financial struggles.

"What is it, sir?" I asked.

"We have $75.00 USD and $2000 JJ as your monthly remuneration," he told me. "JJ" was the local Liberian currency used alongside the U.S. Dollar, with the exchange rate at 15 to 1 USD at the time.

"Okay, I will accept it," I responded without negotiation.

He was shocked but said, "Thanks."

About a year later, I realized my actual remuneration came from the Global Board of General Ministries of the United Methodist Church in New York City, which was $400.00 a month.

Dr. Barcolleh then restricted my practice under the guise of monitoring and evaluating my work. I didn't argue but continued to do my best. Patients began gravitating toward me due to my empathetic approach and patient care skills, where patients became part and parcel of their care and management. When Dr. Barcolleh realized he was losing favor among patients and locals, he began a negative propaganda campaign, insinuating that an investigation was being conducted into the authenticity of my credentials. He would make these general assertions in the hallway, but I remained focused on making a difference in the lives of my patients.

One day, a young lady Dr. Barcolleh had been treating for almost a year with no significant results came to me for consultation. After my initial assessment, I diagnosed her with suspected pyo-salpingitis and advised surgery. She accepted, and on the day of the surgery, while I was prepping, someone informed Dr. Barcolleh. He rushed into the operating room, breaking all the rules regarding surgery.

It was the day he experienced another side of me. I challenged him, saying, "If you make any attempt to disrupt this procedure, it will not be easy between us. I accepted your behavior out of respect, but if it interferes with patient care, you will regret it."

With that, I resumed scrubbing and performed the surgery, finding an abscess within the fallopian tube. The young lady is now healthy and has a child. When all his attempts to undermine me failed, Dr. Barcolleh tried to launch an investigation into the medical school I attended. When the results came back in my favor, he began a clandestine private practice in Kakata, about 150 kilometers from Ganta, often leaving the hospital for a week or more without informing me.

Despite these challenges, I remained dedicated to my work, building relationships and providing the best care possible to my patients. Through resilience and perseverance, I continued to fulfill my dream and make a positive impact on the lives of those I served.

CHAPTER 28

—◆◆◆—

A Life-Changing Moment

On the morning of March 3, 1998, one of my scheduled surgical days, my boss and senior colleague were nowhere to be found. My surgical days were Tuesdays and Thursdays when I typically scheduled a minimum of twelve procedures. I began my day at 6:00 AM by making rounds on the most critical patients and started surgery at 7:00 AM sharp. This Tuesday, I followed my usual routine, completing the first two procedures. As I walked out to my office to see someone, I noticed everyone was looking at me strangely, as if I had done something wrong or they had heard something about me.

When I got to my office, I looked at myself, wondering if I had blood on my scrubs, but everything seemed normal. I called my assistant to find out what was going on.

Jerry came in and said, "Yes, doc, you sent for me."

"Yes, I did," I replied, but quickly followed up with my inquiry. "What is going on? Everyone is behaving a bit odd towards me."

"Oh doc, you don't know? You have been appointed Chief Medical Officer of the hospital," Jerry told me.

Wow, I thought to myself. How did this happen, God? From being a cleaner and janitor to now becoming Chief Medical Officer of the same institution. With that information, I completed my surgeries and held a meeting with the core team in the evening, scheduling a general meeting for the next day.

The general meeting began with praise and worship, thanking God for His continuous blessings and for transforming an ordinary dark stone into a bright, radiant one. Excited and mindful of the nurses' and other staff's dissatisfaction with their remuneration, based on my own experience, I decided to increase everyone's salary by 35%. There was much jubilation. After the meeting, the chief accountant and acting administrator came to me privately and said, "Thanks a million, but chief, where are we going to get the funds to implement the increment?"

"Wow, we don't generate that much?" I asked.

"To be honest, we are struggling to meet the monthly payroll and other liabilities," he informed me.

I thanked him and went to my office, praying for a solution to the mess I had caused. Two days later, I decided to make a trip to Monrovia to contact my sponsors in the United States. At the end of that week, I made the trip and spoke with Mrs. Bettie Story, who then talked with others. They were able to generate enough funds to cover the salary increase, ensuring my promise came to fruition. Fortunately, the staff didn't notice the added stress my action had placed on the management.

That experience reshaped my thinking, making me realize that I needed to include management in my skill set. I understood that my impulsive decision could have caused chaos in my first month of leadership. From then on, I focused on innovation and sustainability, looking for creative ways to generate additional funds for the hospital to manage the wage bill effectively.

I engaged the management of Soquipa Plantation (Oil Palm and Rubber) in Dicke, Guinea, just across the Liberian border. Divine intervention occurred when the wife of the Deputy Manager of the Plant had a medical issue. I diagnosed her condition and prescribed appropriate medication, leading to immediate positive results. Grateful, her husband committed to supplying me with two kilos of beef weekly and purchased me a bed. He kept his promises.

Building on this relationship, the manager approached his top leadership team to enter into a contract where their employees could receive treatment from the clinic run by Ganta Hospital in Diecke. Cases that the clinic couldn't handle would be transferred to the main Ganta Hospital for appropriate treatment. This arrangement brought much-needed funding, relieving the operational budget stress.

As a young health administrator, I focused on morale-building and rebranding the institution. I made a trip to the United States, speaking at churches and the Methodist Hospital in Peoria, Illinois. This trip resulted in two forty-foot containers filled with beds, mattresses, surgical instruments, and other essentials for the hospital, as well as a modern ambulance.

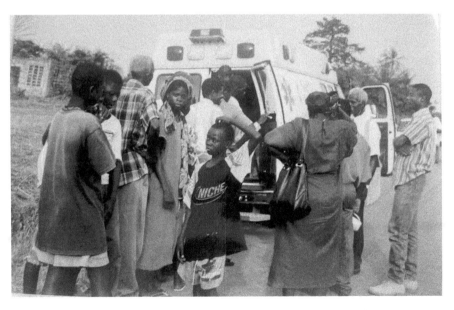

The first modern Ambulance at Ganta Methodist Hospital, provided by the Grace UMC, Bloomington, IL

Not pleased with the existing care, I began recruiting additional staff. I convinced a nurse in Yekepa, Nimba, to join us and recruited another nurse from Danane, Cote d'Ivoire. After running the hospital as the only physician for about two months, I had enough funding to hire more physicians, enhancing the capacity and quality of care.

For the recruited physicians, I ensured their residences had all the basic necessities, such as beds with mattresses, utensils, living room sets, and other amenities. I didn't want them to go through what I did when I first arrived. I even hired a senior colleague, an internist, and paid him more than I received as head of the institution.

With the right team in place, I focused on improving the hospital's image. Using low-cost materials like mud bricks, we enclosed the hospital, leaving four entry and exit points. I negotiated with the Illinois Great River Conference to purchase a mini sawmill for making various sizes of timbers for construction and generating extra funds for the mission station.

Although I wasn't an entrepreneur, my thought process became focused on sustainability. I pinpointed two recruits and offered them scholarships to pursue Bachelor's degrees in nursing at Cuttington University in Suakoko, Bong County. They later graduated with honors, elevating the Winifred J. Harley School of Nursing to a Bachelor's program administered by Liberians.

To maximize opportunities for the hospital's sustainability, I built relationships and partnerships with various United Methodist conferences in the United States, which provided continuous support through an Advance Special. Appreciative of the grace that took me from humble beginnings to a position of influence, I continued finding opportunities for others, striving together to build a health system that met international standards.

CHAPTER 29

———— ◆◆ ◆ ◆◆ ————

Returning to the United States

As Chief Medical Officer of Ganta United Methodist Hospital, it became clear to me that creativity would always put the institution in a better position for sustainability. One of our advantages was the pivotal role we played during the civil crisis, caring for both civilians and combatants injured by bullets. This led to funding from the Leahy War Victims Fund through USAID to offer prosthetics and orthotics to those with injuries resulting in amputations or deformities from the civil war and even polio. USAID contracted UNICEF to manage the project, leading to the construction of a high-tech workshop for manufacturing and fitting prosthetics and orthotics.

However, due to bureaucratic processes, the project timeline lapsed, putting it in jeopardy. To revive the project, the Head of the War Victims Fund, Mr. Lloyd Feinberg, traveled to Liberia via Sierra Leone, where they were undertaking a similar project.

A day before his arrival, Scholastica Kimayo, the Country Representative of UNICEF, called me on the short-wave radio, requesting to see me immediately. I left early that morning from Ganta to Monrovia and was at her office by 8:00 AM.

The UNICEF Representative welcomed me and invited me into her office. She said, "Francis, let me be blunt with you. We are in big trouble of losing the funds. Lloyd Feinberg just came from Sierra Leone, and my colleague told me that he has just ordered the funds for the Sierra Leone project to be halted. The same could happen to Liberia. You are the only one who can save the funds from being halted or continued. That puts me in a very challenging position."

After our pre-discussion meeting, Mr. Feinberg and his team arrived in the conference room of UNICEF. The office was located near the Organization of African Unity Center in Virginia, Montserrado County, just west of Monrovia. Mr. Feinberg was introduced and went directly into business. He started by indicating he had just halted the program in Sierra Leone due to mismanagement of funds. He added, "The same may happen to Liberia if you, Francis, can't convince me why we should continue to pour US citizens' tax money into Liberia." A presentation was done on the project's current status, but he continuously asked questions, indicating that there was certain information he had that would warrant the closure of the project. Not privy to such information, I said to him, "Mr. Feinberg, the project in Liberia came into being because people cared about those who lost their limbs due to the carnage in both Liberia and Sierra Leone. Whatever the situation was that made you cut the funding in Sierra Leone is best known to you, but regarding Liberia, yes, there have been numerous delays due to the bureaucratic process of UNICEF and delays in procurement. However, the equipment is here now, and the building is completed. If you decide to stop the funding, it would significantly impact those who would have become hopeful again. Nonetheless, the hospital would be the winner, and you may not be able to complete the chapter on the Ganta Prosthetic Workshop because you were not willing to give a second chance."

He immediately interrupted. "Is the construction completed, and is the equipment there?"

"Yes indeed, sir," I responded.

He then signaled to his security advisor, requesting clearance for him to go to Ganta. This was a curveball because the building was not fully completed. The structure was done, but the building was only white-washed; the actual painting wasn't done. Despite the restrictions against going beyond Kakata, 50 miles outside of Monrovia, by US citizens, he requested a special exemption to go there at his risk. His request was granted, so he was going to be in Ganta by the latest 12:00 noon the next day, and a decision would be made while in Ganta.

With that information, we had to do the extraordinary. I immediately left the meeting, headed for Ganta, and upon arrival, called a friend of mine, a civil engineer, to gather all the painters in Ganta to come over and have the building painted. That was a herculean task from perception, but as the work began, it seemed doable. The painting was completed by 3:45 AM, leaving all the windows and doors open for ventilation and removing the fresh scent of the paint from the building. By 11:15 AM, the team arrived in Ganta.

Upon arrival, the team participated in a program with speeches and songs. Mr. Feinberg mentioned during his response that he was impressed with the structure and couldn't wait to take a tour of the building to make an informed decision regarding the 2.5 million dollars in funding. At the end of the indoor program, the tour of the Orthopedic, Prosthetics, and Orthotics Workshop began. The tour was very successful, and Mr. Feinberg clearly indicated that he was overly impressed with the level of work done and was excited to sign off on the funding to begin making a difference in the lives of those who lost their limbs during the civil conflict and those victimized by polio and other debilitating diseases.

Mr. Feinberg then called me aside for a one-on-one discussion. He congratulated me on my leadership and ensuring that Liberians received this passionate gift from the people of America through the Leahy War Victims Fund sent through USAID. He then asked if I wanted the new funds to go through UNICEF again, and he clearly indicated due to the previous delays from UNICEF that he wasn't comfortable using that framework any longer. Based on that, I informed him that the funds could go through UMCOR (United Methodist Committee on Relief). He was very excited about this new mechanism for the transfer of funds that would bring relief to the people of Liberia. He then announced that

Liberia was going to receive the funds and the mechanism under which the funds would be received and accounted for would be worked out when he got back to the United States of America. Upon the announcement, the entire campus of Ganta and the city were ignited with joy, knowing that the people of Liberia would be the beneficiaries of America's benevolence.

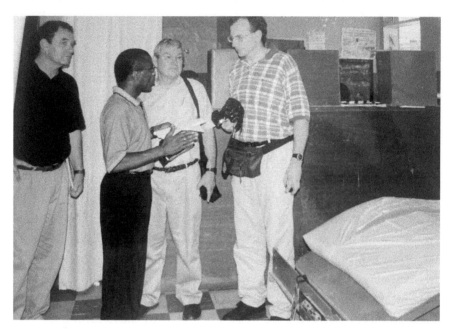

Mr. Lloyd Feinberg, Former Manager of the Leahy War Victims Fund in Black T-shirt

As the team departed, I went directly to my room with tears running down my cheeks. I gave glory to God, who continuously worked through me as an instrument for His work. I consciously made sure that I didn't take His glory from Him. In my moments of solitude, I often paid homage to His grace and what He had done and would continue to do. Like Francis of Assisi, I often reflected on his words, "Lord, make me an instrument," and that is what I wished to be, in whatever form or package His grace allowed.

CHAPTER 30

---❖❖◆❖❖---

The Hypocrisy of My Departure from Ganta Hospital

Board Members of The Ganta United Methodist Hospital: (L-R, Hon Rachel Miller, Bishop Arthur F. Kulah, Hon Yador Saywon (late), Rev. Nyan Gamie Zuagele(late)... Standing Bro. J. Lamax Cox (late) Rev. Dr. Emmanuel Bailey, Rev Anna Kpan (late) Dr. Kateh, Dr. Jewell, Dr. Sei Parwon(late), Saye Glayenneh, and Mr. Miayen (late)

The Ganta United Methodist Hospital's board members gathered to celebrate the approval of the Prosthetics and Orthotics Workshop. The approval had spread widely around Nimba County, Liberia, and neighboring countries, but beneath the celebration, a plan of hypocrisy was unfolding. It all started with the bureaucratic hurdles that almost jeopardized the project. I had recommended the United Methodist Committee on Relief (UMCOR) to manage the funds, hoping to avoid those delays. But I was wrong.

Once the funding was approved, UMCOR's head had a plan to use a significant portion of the grant for administrative costs, compromising the amount meant for the victims. At a historic board meeting held in Ganta in mid-October 2000, a UMCOR representative informed the board of the 2.5 million dollars they received and outlined the administration plans without fully explaining the implications. As the board secretary, I raised concerns about the overhead costs, which were rumored to be over twenty-five percent of the total grant. When the representative couldn't provide a clear answer, the board voted not to approve the grant's implementation until we received clarification from Rev. Dr. Paul Diedak, head of UMCOR.

The Liberia Annual Conference was on the verge of electing a new bishop, as Rev. Dr. Arthur F. Kulah's tenure was ending in December 2000. The West African Central Conference was scheduled to nominate a new bishop. Rev. Dr. Kulah, a father figure to me, had played a crucial role in my journey to becoming a medical doctor. He had appointed me as Chief Medical Officer of Ganta Hospital, defending my position against a missionary who tried to undermine my role. The election of the new bishop would significantly impact the hospital and the grant implementation.

Two candidates were approved for the election: Rev. Dr. Julius Sawolo Nelson Jr. and Rev. Dr. John Innis. I feared that if Rev. Dr. Innis won, he would compromise the hospital's future to satisfy his former employer, UMCOR. Unfortunately, Rev. Dr. John Innis was elected. He implemented a plan crafted by Rev. Dr. Paul Diedak to ensure I was removed from my position to eliminate any hindrances to their plans.

The first Annual Conference presided over by Bishop Innis was held in Kakata, Margibi County, on the campus of Booker T. Washington School. As usual, I ran a mobile clinic at the conference site. In 2000, the first modern ambulance donated to Ganta United Methodist Hospital by Grace United Methodist Church in Bloomington, Illinois, was used as the clinic. Many attendees from rural Liberia had the opportunity for physical checkups, and those needing further treatment were referred to Ganta United Methodist Hospital. The ambulance attracted much attention, adding flair to the conference.

To execute their plan, I was called out of the assembly during a session and brought to the podium. Bishop Innis announced that due to my excellent service, I was awarded a scholarship for intensive training to prepare missionary doctors for work in West Africa. Rev. Dr. S. R. E. Dixon prayed for me, and although shocked, I did not resist. It felt like I was being led to a sacrificial altar. After the prayer, I wondered what would come next.

After the conference, I returned to Monrovia. Two days later, I visited the Bishop's residence in Congo Town to discuss the scholarship details. In his bedroom, with Rev. Dr. James Karblee present, I asked the Bishop about the training and sponsorship.

The Bishop said, "Dr. Kateh, you have played a pivotal role in my election. I will not deceive you or let you down. The details are being worked out by UMCOR through Dr. Cherian Thomas."

I asked if the training could be arranged in an African country like South Africa or Kenya. The Bishop insisted it should be in the United States to maintain confidence in the training. I accepted and followed the steps to get a one-way ticket to the US.

Leaving the Bishop's residence, I felt a mix of sadness and hope. I informed my wife, who was ambivalent about the situation. "What will we do? This is sad, and I am feeling very down," she said.

I reassured her, "God, who brought us this far, will take care of us." Deep down, I was afraid of the unknown future. I discussed the situation with my American guardians, who promised to pick me up from O'Hare International Airport in Chicago.

On April 28, 2001, an elaborate farewell program was held at Ganta United Methodist Hospital. Tears were shed by staff, community members, and others. It was one of the saddest days of my life. Working at Ganta Hospital had been the pinnacle of my career, and leaving without accomplishing all I had planned felt like an unfinished chapter.

CHAPTER 31

―――・◆◆ ◆ ◆◆・――――

The Die Was Cast

On May 4, 2001, I departed Liberia and landed at O'Hare International Airport in Chicago, Illinois. My guardian, John Doyle, now deceased, had arranged for my transportation on a Peoria Chartered Bus from Chicago to Peoria, Illinois, where he met me upon arrival and took me home to Henry, Illinois. "Dad and Mom," as I affectionately called them, had recently built a new home in Henry, leaving their long-time farm residence to a new family, with the land remaining for farming under the care of my older brother, Ron. The new house had three bedrooms, and I settled into one of them, feeling both comforted and anxious about my future.

I spent over six weeks in a state of limbo, waiting for information about the supposed "scholarship" that had brought me back to the United States. All attempts to discuss the scholarship details with Dr. Cherian Thomas, who was supposed to provide the necessary information, led nowhere. Finally, a contract arrived in the mail, and when I opened it, I was shocked by its terms:

- I had requested a five-year leave of absence.
- During this period, I would have no lien on the hospital, couldn't speak on behalf of the hospital, nor raise any funds for it.

- The hospital board resolved that I would receive a one-way ticket, $500 for three months, and would need to reapply after five years if I wished to return.

I was devastated. My entire life had been anchored in the church, and I had vowed to serve to the best of my ability. Now, the very institution I trusted had betrayed me, leaving me feeling helpless and betrayed.

"Why me?" I asked myself repeatedly. The answer seemed clear—I had stood against UMCOR's plan to misappropriate over twenty-five percent of the grant meant for victims of the civil war and those affected by polio. Dr. Paul Diedak, using the influence of both my Bishop and Dr. Thomas, had orchestrated my removal from Ganta United Methodist Hospital.

Reading the contract, I lost all appetite and felt nauseous. I retreated to my room, tears streaming down my face. My dad, John Doyle, entered and, seeing my distress, asked what was wrong. When he read the contract, he was furious.

"Francis! What the hell is going on? Why are you crying?" he demanded. After quickly reading the contract, he shouted, "Is this what the Church has come to? Dishonesty and hypocrisy!" He informed my mom, and they both came back to my room, concerned.

"Francis, it will be well," Mom reassured me gently.

Dad, still angry, said, "There are two things you can do: either take the United Methodist Church of Liberia to court or leave them with their conscience. Remember, I will always be here for you." He then left for the dining table with Mom following. I had no appetite and couldn't eat dinner.

The entire night was sleepless as I questioned where I had gone wrong. I felt I had been doing the right thing by ensuring the well-being of the people of Liberia through the Church, but it seemed I had been mistaken. In my solitude, I reflected on the story of Joseph and prayed that God would use me for His purposes. The hymn "It Is Well" echoed in my mind, particularly the stanza, "Though Satan should buffet, though trials should come, let this blest assurance control: that Christ has regarded my helpless estate, and has shed His own blood for my soul." With these words, I decided to leave the United Methodist Church, Liberia Annual Conference, presided over by Rev. Dr. John G. Innis.

I informed Dad of my decision to leave the Church. His eyes red with emotion, he said, "Francis, you spent so much time in Chicago, and I know you love it there. Go to Chicago and find a place. I will pay the rent for a year. We will see how it goes from there."

My heart filled with joy. "Dad, thank you so much," I said.

Later that afternoon, Dad called me to the living room and told me he had spoken to some friends in Chicago about finding a place for me. He had received a call from one who found two places in Hyde Park, Chicago. The weekend passed uneventfully, and at the end of the week, Dad took me to Chicago to look at the apartments. The first wasn't suitable, but the second, located at 1200 S. Blackstone Ave, was perfect. I thanked God for the blessing.

I moved into the apartment in July 2001. The master bedroom overlooked Lake Michigan, and having my independence was a significant achievement, thanks to the Doyle family, who had supported me from college to medical school, to my return to Liberia, and now, my resettlement in the United States.

Despite spending many years in the United States, I had never attempted to change my immigration status. Now, faced with uncertainty, I needed to legalize my stay to work. I sought help from Hartland Alliance, an organization in Chicago that assists immigrants and champions human dignity. Their pro-bono services helped me change my status, enabling me to get a job at Sacred Heart Hospital on Franklin Boulevard in Chicago, IL.

I worked in the Surgical Unit under my mentor, Dr. Jose Deleon. Despite his success, I noticed he often seemed unhappy and depressed. When I asked him about it, he confessed he regretted not pursuing a career in health administration, which would have allowed him more time with his family and a less stressful life. Inspired by his words, I enrolled in the Master of Health Administration program at Governors State University, University Park, Illinois, in 2002. My previous experience as Chief Medical Officer at Ganta United Methodist Hospital had prepared me well, and I graduated in 2004.

Shortly before graduation, I was offered a position as Public Health Director of Anson County, North Carolina, becoming the first African Public Health Director in the state. I started on May 3, 2004.

Three days into the job, my administrative assistant, Brenda, resigned, stating she was uncomfortable working with me. I thanked her for her honesty and wished her well. The Board of Health Chairperson, Dr Richard Constantian, later insisted I rehire Brenda as a consultant to work on the budget. I refused, standing my ground.

From 2004 until I left the Anson County Health Department, the budget gradually increased. I consistently engaged the community on various health matters, addressing issues that could have both positive and adverse effects on the general public.

The job was progressing well, and with ample leisure time after work as a Public Health Director, I decided to apply as an adjunct at Pfeiffer University's Master of Health Administration Program, teaching Medical Jurisprudence and Comparative Health Analysis. This experience expanded my horizon in effective and efficient management and analysis of healthcare systems. As an adjunct, I felt it was important to share my thoughts on critical health issues of the time. Consequently, I contributed to the Anson Record, a local newspaper, with an analysis titled

"Healthcare Reform: Why Should You Care?" Below are a few of my published thoughts on "Obama's Care":

Healthcare Reform: Why Should You Care?

My greatest asset in Anson County was Mrs. Carol Gibson. When I was recruited, she was the Acting County Health Director, and she understood the inner workings of both the County and the State. Many times, I consulted her before making major decisions. I valued her advice immensely, but there was one instance where I didn't heed her warning, and I paid a heavy price for it. She had cautioned me about working with a particular staff member who was not up to the task. I became empathetic and sometimes approved her excuses, which I later realized were fabricated. When she could no longer find excuses, she filed a discriminatory lawsuit against me and the county, which was eventually thrown out of court. Carol was brutally honest and never minced words. My success in the County was due to the bond we had, and I will always appreciate the role she played in my life.

My days, months, and years at the Anson County Health Department were great. I had a fantastic team to work with at the Department, especially the County Manager and other staff. I remember one afternoon when my animal officer came to my office and asked if I ate deer meat. I told him yes, and the next evening he came to my house with two big deer. I told the officer I couldn't eat one entire deer, let alone two. His response was, "I don't know what to do with the other one, so you manage it." I called some friends from Charlotte, NC, and they came over to help me butcher the two animals. We had deer meat for half a year. The following year, Rob came to my office and asked, "Dr. Francis, do you eat wild hog?" I was a bit hesitant, and he said, "I have one for you in the back of my pickup." This is how caring people were, always making sure that my well-being was prioritized. One of my staff, Rennie, would have my two girls over at her place when I had to travel. In short, the people of the County were accommodating and treated my kids and me as family.

Within the community, I became part of the First United Methodist Church and played an active role in its ministries. Many good people at the Church welcomed us wholeheartedly. My kids served as acolytes on many Sundays, and they were so excited that they couldn't wait for their turn to come. Additionally, I became a member of the Rotary Club, which was a great place to meet many community leaders and stakeholders. It was an excellent gathering to know the who's who in the County. It was a difficult decision to join the Rotary and not the Lions, as both invited me to join, but I had to make the right decision, which I never regretted.

CHAPTER 32

——·◆◆◆◆◆·——

Protection of Lives in Anson County

Over a year had passed since my appointment as Public Health Director of Anson County. The devastation caused by Hurricane Katrina in 2005 brought another opportunity into my life. I have always believed that I am an instrument meant to be used positively for the well-being of others. As I watched the scenes of Hurricane Katrina unfold on television, seeing individuals carrying their belongings on their heads and others awaiting rescue on rooftops or interstates, I was reminded of the Liberian Crisis. It made me wonder if I had the expertise to prevent such loss of lives should a similar disaster strike Anson County.

To answer that question, I decided to pursue a degree in Homeland Security. My search led me to the University of Connecticut, Storrs, where I enrolled in 2006. I graduated in May 2008 with a Master of Professional Studies in Homeland Security Leadership, with an emphasis on Public Health Disaster Preparedness.

An article published by The Perspective online newspaper highlighted this achievement:

Anson County Health Department's Director, Kateh, receives UConn degree

The Perspective
Atlanta, Georgia
Posted May 23, 2008

(May 20, 2008) – Dr. Francis N. Kateh, Director of the Anson County Health Department, was presented with a Master of Professional Studies in Homeland Security Leadership degree with an emphasis in Public Health Disaster Preparedness earlier this month from the University of Connecticut.

"The university's homeland security program offers concentrations in transportation, security, water, human resource management, and public health," said Dr. Kateh. "After Hurricane Katrina, I thought I should study homeland security in the area of public health."

"My concentration in public health qualifies me to prepare the public for a disaster, whether natural or manmade, such as terrorism or bioterrorism. I can determine how credible the security is if there is a threat, such as an anthrax incident, and what the best way is to assure the safety of the people."

Dr. Kateh, who has been health director since May 2004, began his studies at UConn in January 2006. "At the beginning of each semester, I spent 10 days on campus to go over the curriculum for the remainder of the semester," he said, adding, "Then I worked online toward my degree." Prior to becoming the Director of Anson County Health Department, Dr. Kateh served as the Medical Director of Ganta United Methodist Hospital in Liberia, West Africa. Dr. Kateh, who has earned a Bachelor of Science Degree, Doctor of Medicine, and Master of Health Administration, claimed that this Master of Professional Studies in Homeland Security is the end for now.

***

My family and I drove from Wadesboro, North Carolina, to Storrs, Connecticut, for my graduation. It was a tiring journey because my wife had graduated two days earlier with a bachelor's in social work, and we had to leave the next morning for my ceremony. As I drove, we were about to cross from New Jersey into New York City when my daughter Frances said, "Dad, the next time you go back to school, I will ask the police to have you arrested."

"Why?" I asked, surprised.

"Because it will be considered abuse to education," she replied.

Her words hit me hard. I realized the internal pain my family was enduring due to my relentless pursuit of education. That statement from my daughter dampened my ambition to become a lawyer, a dream that still haunts me today. However, I have been privileged to study health law as the Public Health Director of Anson County through the University of North Carolina School of Government, which was a requirement for my role. Refreshment courses were required annually, helping me fulfill part of my internal desire.

Through this journey, I learned the importance of balancing professional aspirations with family life. The drive for continuous improvement in public health and disaster preparedness has always been my passion, but my family's support and well-being remain paramount. The experience deepened my commitment to protecting lives and preparing for any crisis that may come our way in Anson County.

CHAPTER 33

———·◆◆ ◆ ◆◆·———

The First Major Experience of My Late Mother's Reaction

When I returned to the United States in 2001 due to the mistrust I had in the Church, my plan was to remain in the US until retirement. I tried to get my father to join me in the United States, but to no avail. In 2009, I decided to visit him and spend Christmas together. During my visit, I stayed at my cousin's house, who was kind enough to make my stay very enjoyable.

During my visit, my father came to Monrovia, and I asked him what happened to his passport. He laughed and said, "My son, I have three questions for you." I said, "Go ahead, Dad."

He asked, "Are there Kru people around the area you live?"

I responded, "No."

"Are there people my age around where you live for me to visit in the morning?" he continued.

I replied, "Dad, I've lived there for almost five years, and I don't even know my neighbors' names. We only wave at each other in the mornings or afternoons when leaving our respective yards. I've never visited them, and they haven't visited me."

"So why do you want me in the United States to sit all day and watch television? My son, thank you for what you are doing for me, but I think I need to be here in Liberia," he concluded.

While in Liberia, I visited my former principal from high school, Sister Mary Laurene Brown. I had always wondered why I was appointed the timekeeper of Our Lady of Fatima High School. After over thirty years, I was finally going to find out.

I booked an appointment to meet Sister Laurene. When we met, after the greetings and other preliminary discussions, I asked her, "Sister, why did you make me the timekeeper at Fatima?"

In a soft voice, she asked, "How far have you gone in education?"

"Sister, I have obtained a Doctor of Medicine, a Master of Health Administration, and a Master in Homeland Security with an emphasis in Disaster Preparedness," I responded.

She smiled and said, "Excellent, this was what I envisaged. I saw a smart young man from a village sitting in class with other students whose parents were influential and wealthy. I did not want the tendency of inferiority to set in. Giving you the bell was the gavel of authority needed to help you concentrate on your studies instead of paying attention to those in the class with you and what they brought to class or how they got to class. Today, you have proven me right, and I am proud of you."

As she explained, tears welled up in my eyes. I wondered how she knew and was able to help me succeed. I thanked her and assured her I would always be grateful and make her proud.

When I returned to the United States, I rested for two days before going back to work on Monday morning. As part of my usual routine, I thanked God for another opportunity to serve Him before starting my day. But that morning felt different. I sat behind my desk, and as soon as I lifted my head, my eyes were fixated on my mother's photo. This photo, taken on Christmas of 1988, was the last picture I had with her and the only one I owned. During the many years of isolation, it was my most valuable asset and motivator. The photo always had a special place wherever I worked.

That morning, as I stared at the photo, I began to weep. Within a few minutes, my administrative assistant walked in and said, "Your guest is here." When I lifted my head, she saw my tears and asked, "What is going on?"

"I don't know," I replied, but the tears kept flowing.

She said, "I think you need to go home. It seems you are homesick. I'll reschedule the guest."

I left through the back door, got in my car, and went home. When I arrived, I went directly to bed because I had a slight headache. I can't remember how long I slept, but I had a dream that felt very real. I found myself in Karlokeh on the farm with my mom, telling her, "Mom, when I grow up, I will be a doctor and will go from village to village and help others." I woke up and thought, "This is very weird." It was the third time I had dreamt about her since her passing.

When my wife came home, I told her about the dream. She said, "I hope you're not homesick."

The next day, I went back to work, and everything returned to normal. I continued fulfilling my responsibilities as Public Health Director, ensuring that the Department had the requisite manpower to make a positive impact on the well-being of the people I served.

Health Director seeks more staff

BY THOMAS LARK
The Anson Record

The Anson County Health Department needs more funding and staff.

That's how Dr. Francis Kateh described his department's situation to the Anson County Board of Commissioners at its April 3 regular meeting.

"We're trying to change the perception of the Health Department as a place that serves indigents," Kateh said. "We want to maximize our services with a minimum of staff. But that can mean a burn-out of staff."

A shortage of three nurses, a health educator and a social worker for various programs has caused a need for others to step in, said Kateh.

"We need assistance from you commissioners to help us devise a system that will enable us to recruit and retain staff, most especially nurses," he said.

Between July 1, 2006, through Jan. 31, Kateh said, the Health Department saw 454 adult clients, of whom more than half—51 percent—paid nothing for their visits, while 140 were Medicare, Medicaid or private insurance clients.

Kateh also said the department's flu season walk-in flu shot clinic vacinated more than 300 people in a single day.

Kateh said he'd like to see more County employees take advantage of Health Department services.

Kateh also gave printed versions of his department's annual report to each commissioner and also provided a copy to *The Anson Record*. Among the 2006 statistics the report states are the following case numbers for Anson County:

- AIDS—three
- HIV (reportable)—nine
- syphilis—four
- gonorrhea—91
- chlamydia—134
- hepatitis B (acute)—one
- tuberculosis—one
- salmonella—seven

Kateh thanked the commissioners for any assistance they could provide.

"It is my ardent hope that we can begin this dialogue now," he said, "so as to avert an imminent disaster in the future."

CHAPTER 34

————— ◆◆ ◆ ◆◆ —————

The Need for a Change

After serving Anson County in North Carolina for over five years, I felt it was time to move on. When I applied for the job in Anson County, I had told them that I would stay a little more than three years; true to my word, I had spent over five years there. It was time for a fresh start. As a result, I began searching for a more competitive and challenging job.

The first place I applied was the State of Georgia, which had a vacancy for the position of State Health Director. I applied and went through the interview process. Here are excerpts from communications between the HR Director and me:

From: Andrea Fuller-Ruffin
To: Frankateh

Dr. Medows and Dr. Edwards asked me to schedule conference calls with senior leadership in the Divisions of Public Health and Emergency Preparedness and Response. I have asked the three staff to provide some available times for me to schedule a conference call with you. Deborah Bevelle has indicated that you will be unavailable between Sept. 25 and Oct. 2. I will contact you soon with some possible dates and times that we can coordinate with your schedule.

If you have any questions in the interim, please feel free to contact me.

Andrea Fuller-Ruffin, Director
Office of Human Resources
Ga Department of Community Health

████████████████████████

From: Andrea Fuller-Ruffin ████████████████████
To: Frankateh ██████████

It was a pleasure talking with you yesterday. I hope you had a safe trip.

I have been able to obtain some dates for the week of Oct. 5th, but the only date that Tom Wade, Miriam Bell, and Pat O'Neal are all available is Oct. 8th. I can schedule you to meet with Tom and Miriam at 1:00 and with Pat at 2:00. Please confirm your availability to meet with them in person on that date. Thank you.

Andrea Fuller-Ruffin, Director
Office of Human Resources
Ga Department of Community Health

████████████████████████

From: "Andrea Fuller-Ruffin" ████████████████
Date: Fri, 2 Oct 2009 12:41:06 -0400
To: <Frankateh ██████████

Subject: DCH Background Checks

As part of the hiring process, DCH verifies educational credentials, checks references, and conducts criminal and financial background checks. The verification of the educational credentials and reference checks are handled internally by the DCH Office of Human Resources, and the criminal background check is conducted by the DCH Office of Inspector General. If you are the top candidate, we will have the financial background checks completed by the Georgia Bureau of Investigation (GBI). I will notify you at that point in the process.

I am initiating the education, reference, and criminal background checks. I need for you to provide at least two (2) professional references. These should not be peers but individuals who supervised you or had oversight over projects. I am attaching the forms that need to be completed and returned to me by fax, by mail, or when you come to meet with PH leadership on October 8th.

If you have any questions, please feel free to contact me.

Andrea Fuller-Ruffin, Director
Office of Human Resources
Ga Department of Community Health

(frankateh ▮▮▮▮▮▮: you (Bcc) + 1 more
Mrs. Fuller-Ruffin,

It was so good meeting with you yesterday. I am internally grateful to you for taking up your time to make sure that I was at the right place for my meetings.

As promised, please find as an attachment the Salary History and Approval for Reference Check.

Thanks,
Dr. Francis N. Kateh

After completing the reference checks, I received a call from the HR Director, Ms. Fuller-Ruffin. She inquired, "When did you become a US citizen?"

I told her, "No, I am not, but I am a Permanent Resident."

She exclaimed, "Oh my God, I will contact you later." Later, I was informed that since the position required restricting movements and quarantine in case of an outbreak, it was difficult to appoint someone to the position who was not a citizen. I thanked her for helping me through the process.

Undeterred, I continued my search and eventually found another vacancy in Tompkins County, New York, for the position of Public Health Director. The process was challenging and very competitive. Here are excerpts from our correspondences:

Original Message

From: Stephen Estes ██████████████████
Sent: Thu, Apr 1, 2010 6:28 pm

Subject: Public Health Director - Tompkins County, NY

Dear Candidate:

Congratulations! You have been selected to move forward in the process with regard to our Public Health Director search. The interview team has provided three questions for your consideration and thoughtful response. They have requested that you answer all three questions in no more than three pages total (not three pages per question). It would be greatly appreciated if your responses were in Microsoft Word format, 12-point type, single-spaced, with one-inch margins. Please send this document to me as an e-mail attachment no later than 4:30 PM, Monday, April 12, 2010. If you have a conflict that would preclude you from being able to provide responses by that date, please let me know as soon as possible. I will share your reason with the team, and they will determine whether or not an accommodation will be made. The questions are as follows:

1. Please share your vision of public health now and in the future.
2. Given the economic climate in New York State, what do you see as the greatest challenges facing county health departments in the next 5 years? What steps would you propose to meet these challenges?
3. The Public Health Director supervises a staff of 80 and is responsible to collaborate with educational, medical, and community constituents. In addition, the position must balance autonomy, accountability, and authority to varying degrees with the Legislature, NYS Health Department, County Administrator, and the Board of Health. How does a Public Health Director balance the competing requirements of employee groups, community constituents, regulatory agencies, etc.?

Please reply to confirm that you have received this e-mail. Your responses to the questions above will guide us with regard to who moves forward with a first-round interview. Again, I would like to thank you for your interest in employment with Tompkins County and for the effort that you have already invested in this process. We look forward to reviewing your responses. First-round Interviews will happen April 30th, May 3rd or May 6th. I will be in contact with you again on April 21st to let your know your status and, if you are moving forward, to schedule your interview. Once more, I wish you the best of luck as the process unfolds.

Sincerely,
Stephen

Original Message
From: Stephen Estes
To: frankateh
Sent: Mon, Apr 12, 2010, 2:13 pm

Subject: Re: Public Health Director - Tompkins County NY

Good afternoon Dr. Kateh:

Thank you for the questionnaire responses. I will make sure that the interview team reviews your materials. In answer to your question, the timeline has been set and first-round interviews will happen across three days: Friday, April 30th, Monday, May 3rd and Thursday, May 6th. Finalist interviews will happen May 17th, so it appears as though you do have a conflict. The Interview Team is reviewing questionnaire responses all this week and will meet next Tuesday, April 20th to decide who will move forward to the first-round of interviews. I will mark my calendar to send you an email next Tuesday evening (after the meeting) or Wednesday morning to let you know where you stand in the process. I can share the conflict with them now if you wish but am unsure what we could do to accommodate you as we need to get this vital position filled and it is not practical to stretch the recruitment out further.

Stephen

Mr. Sestes,

Thanks for the expedited response. For now, there is nothing I can do or say until the revision of the responses are completed. If you can recall, you asked me to send a letter indicating when I would be out of the country and the return date, which was done, because I indicated to you my commitment on a work team. If I were aware of the inflexibility of the schedule, perhaps I would not have gone through with the rigorous exercise to compete the questions. Whatever the outcome is, I am grateful that I was able to go through the first phase of the process.

Thanks again.
Respectfully yours,
Dr. Kateh

Stephen Estes ███████████████████████

Dear Dr. Kateh

Once again, congratulations on moving forward to the finalist phase of our interview process. The finalist interview will consist of two components: a twenty-minute Microsoft Powerpoint presentation and about an hour of questions. You will have approximately 30 minutes to meet Legislators and Health Department staff at the Health Department immediately prior to your interview. Please show up at the Personnel Department, 125 E. Court St., at about 12:45 PM, and someone will escort you to the Health Department and bring you back for your 2:00 interview.

The team has settled on "Vision" as the topic of the Powerpoint presentation. You must demonstrate your vision of public health in Tompkins County; where are we now and where we need to go. The Powerpoint presentation should last no longer than twenty minutes. Please send me a copy as an email attachment as soon as possible and I will test it on this end to make sure our technology is compatible. Let me know if you have questions. As always, best of luck as this process unfolds.

Sincerely,
Stephen

Good afternoon Dr. Kateh:

I would like to congratulate you on being selected as a finalist candidate for the Tompkins County Public Health Director position. We had some difficulties with the conference call, but the Team was very impressed with your responses to the questions, as well as the aplomb with which you handled being cut off three different times.

I realize that you are coming back into the United States on May 16th. Finalist interviews are scheduled for Monday, May 17th. I have held you a spot for you later in the day at 2:00 PM. I would like to speak with you this coming Monday morning via telephone to gage your interest and discuss the logistics of getting you here on May 17th. I attempted to call your prepaid cell phone but got no answer. Please call me at 9:00 AM (roughly 2:00 PM your time) this coming Monday to discuss. ███████████████████████.

Though I cannot pay for your airfare from Liberia, I am authorized to reimburse you the amount of a round-trip airfare to and from Wadesboro, NC. In addition, I can reimburse you for the cost of one night in a hotel should you wish to fly in on Sunday prior to the Monday interview. Meals and ground travel expenses will be covered as well.

Please give me a call Monday, May 10th, at 9:00 AM to discuss in detail. Thank you.

Stephen

To be honest, I could not understand how Mr. Estes was capable of working out the logistics for me to participate in the process, but it all went well and every time, the results were in the affirmative and the process was seamless. After the final interview that took almost the entire day, I got to the hotel to rest, and a call came after couple of hours form Mr. Stephen Estes regarding the Committee's decision. As usual, I was excited but give God the glory since I believed He is the master-crafter of my destiny. Below was an email from Mr. Estes:

Good day Francis:

Again, congratulations on the offer of employment. I was pleased to hear that you are definitely interested in accepting. It was also very nice to spend a few hours with you yesterday evening, getting to know you and showing you around the Ithaca area. If you have any questions at all, please feel free to give me a call. If your wife and sons wish to take some time to check out the area, I am happy to help in any way I can; whether it is taking you on a tour of the area, connecting you with a realtor or rental agent, arranging for a tour of the primary and middle schools, facilitating community connections, etc. Just let me know what you need.

Commissioner Fitzpatrick will be in contact with you on Friday via your cell phone number to discuss specifics. However, I do need to follow up on a few things. As you know from the job posting and our discussions, appointment to the position of Public Health Director is made by the County Administrator. However, appointment is also contingent on confirmation by both the local and State Boards of Health and is also subject to confirmation by our local legislative body. Based on your qualifications, I do not anticipate a problem with this process, however, I do need to gather documentation to present to these entities that will support your stated qualifications.

With regard to your Masters in Health Administration Degree, I will need you to contact Governor's State University and have them send an official college transcript directly to me as soon as possible.

I would also like to have an official copy of your University of Connecticut Homeland Security Leadership MPS/HSL transcript on file. Again, have them send the transcript directly to me. My address is in the footer of this email.

Please provide proof of citizenship, passport and naturalization paperwork.

It would also be nice if I could get some clear copies of the certificates that you submitted with your initial application. Many of them did not come through cleanly. Please feel free to submit any other documentation that you believe would strengthen your position before the board of health. Thank you, and I very much look forward to working with you in the future.

Sincerely,
Stephen

Anita Fitzpatrick ████████████████████████

To: you Details

The process of trying to obtain special legislation to exempt citizenship for a three-year period is still under way in Albany. No promise yet, but it does look promising. I will update you as soon as I hear back - perhaps Monday.

Alice Cole inquired about your phone call. At this point, she will wait to call you back until the exemption is resolved and the job is offered. When Alice retires, she plans to remain in the area, so will be reachable. The County Administrator will likely announce an interim director early next week, as Alice Cole will be retiring by Wednesday.

In conducting a further background check, I have a question about the criminal history section of the report:

The report lists two arrests, on or about 7/9/2007 and 1/15/2008, both misdemeanor level arrests and 1 misdemeanor conviction (2008). Can you tell me if the arrest record is accurate and if you have a misdemeanor conviction. It may also be an error in the report, but I do have to follow up on this issue with you. Your cooperation in getting additional information is greatly appreciated.

I will be in touch next week as soon as I have an update.

Letter above

Stephen Estes ███████████████████

To: you + 1 more Details
Good afternoon Dr. Kateh:

On behalf of Joe Mareane, I am writing to give you an update on the status of the pending legislation. The bill was delivered to Governor Patterson's office on July 19th. The Governor has until July 29th to either sign the bill into law or not. I would expect that Joe will contact the Governor's office on Friday the 30th and either he or Anita will be in touch with you shortly thereafter to discuss the outcome of that effort. As always, I wish you the best of luck. Sincerely,

Stephen

Original Message
From: Anita Fitzpatrick ██████████████████
To: Frankateh ████████████
Sent: Mon, Aug 2, 2010 5:35 pm

Subject: Update - Public Health Director

Hello Dr. Kateh

I am pleased to report that the Governor has signed the special legislation regarding citizenship requirements upon appointment. The legislation allows for a three (3) year waiver to allow you to complete the citizenship process.

The remaining steps include:

Confirmation by the Board of Health (8/10 - anticipated)
    Discussion of terms and conditions (Salary, benefits, anticipated start date), as well as how we can best assist you and your family to transition successfully.

Confirmation by the Legislature

Please let me know if you have any questions or how it would be most helpful to move forward with preparing for your appointment. We are very pleased that you will be joining the organization and assuming this pivotal role in public health for the citizens of Tompkins County.

The process of seeking a new position was both daunting and invigorating. Each step I took toward a new opportunity reminded me of the many transitions and challenges I had faced and overcome. Moving from Liberia to the United States, navigating through different roles and responsibilities, and continuously striving to improve my skills and qualifications—all these experiences had prepared me for the next chapter.

In my heart, I knew that the journey ahead would be just as demanding, but the prospect of making a difference in a new community kept me motivated. The desire to bring my expertise to new settings and to continue serving those in need was a driving force that pushed me forward.

As I awaited responses and further steps in the hiring process, I reflected on my time in Anson County. The relationships I had built, the lives I had impacted, and the lessons I had learned were invaluable. I felt a deep sense of gratitude for the experiences that had shaped me and prepared me for whatever lay ahead.

While the future was uncertain, I embraced the possibilities with hope and determination. The need for change was not just a professional aspiration; it was a personal mission to continue growing, learning, and contributing to the well-being of others.

With each application and interview, I felt a renewed sense of purpose. I was ready for the challenges that awaited me and confident that I would find the right opportunity to make a meaningful impact once again.

CHAPTER 35

———·◆◆ ◆ ◆◆·———

Trip to Ithaca, New York

My wife, the kids, and I traveled to Ithaca, anticipating a potential offer after being confidentially informed that the process was ninety-five percent complete, pending the Governor's approval. Deputy Commissioner Stephen Estes warmly welcomed us, introducing my family to realtors and various elementary and middle schools in different subdivisions.

After a tiring day and a delicious meal at a local restaurant recommended by Stephen, we retired to the Holiday Express Hotel. Around 4:00 AM, I was abruptly awakened by my wife shaking me. Tears were streaming down my face. I quickly realized I had been dreaming.

"What's going on?" my wife asked, concerned.

I explained, "I had a dream where my mother was very unhappy with me. It felt so real. We were in Karlokeh, on the farm, weeding the rice fields. I was telling her about my plans to become a doctor and help people in the villages. But when I looked up, she was angry. I begged her for forgiveness, but she wouldn't forgive me. That's when you woke me up."

My wife tried to reassure me, saying it was just a dream. But deep down, I felt it was more than that.

After getting ready in the morning, we headed to the cafeteria. As soon as we stepped out of the elevator, my phone rang. It was from the office of the President of Liberia, Madame Ellen Johnson Sirleaf.

"Can I speak with Dr. Kateh?" she asked.

"This is he," I replied.

"Could you be my guest at the July 26, 2010, Independence Celebration?" she asked.

"Yes, Your Excellency," I responded. She then passed the phone to Mrs. Toles to get my information for processing the travel arrangements.

When I got off the phone, I was puzzled. Why would the President of Liberia invite me to the Independence Day celebrations? My wife, curious, asked who I had been talking to.

"It was the President of our country," I said.

"You mean President Sirleaf?" she asked in disbelief.

"Yes."

"What does she want from you?" she inquired.

"Nothing, she just wants me to be part of the Independence Day celebrations."

My kids were thrilled. "Dad, will you sit close to the President? If you do, take a picture with her, then you have our approval to go."

I promised them I would do my best to deliver. Although my wife wasn't too happy, the kids' excitement and approval made me feel better.

The rest of our time in Ithaca went well. We identified a few houses for rent with an option to purchase and found suitable schools for the kids. That evening, one of the kids suggested a walk since the hotel was near Lake Ithaca. We all agreed, turning it into a reflective stroll, absorbing the day's events, and discussing our potential future in Ithaca.

During the walk, my wife and I talked about what life could be like in Ithaca. I admired the place, especially because of Cornell University. The prospect of becoming a Public Health Director and possibly teaching at one of my favorite universities filled me with excitement and anticipation.

The walk was refreshing, and it prepared us for a good night's rest before driving back to Wadesboro, North Carolina, the next morning.

CHAPTER 36

———·◆◆◆◆◆·———

Answering the Call

Upon returning to Wadesboro, North Carolina, I began preparing for my trip to Liberia. I left Wadesboro on July 20, 2010, and arrived in Monrovia on the 21st. My cousin met me at the airport, insisting I stay at his house instead of a hotel. That evening, I called the contact person from the President's office to let Her Excellency know I had arrived.

The next morning, my cousin woke me up with news that Dr. Bernice Dahn, the Chief Medical Officer of Liberia, was on the phone. She welcomed me back and mentioned that Minister Dr. Walter Gwenigale wanted to meet with me and others at the Chinese Embassy for dinner. I thanked her but admitted I didn't know where the Embassy was. She laughed and said my cousin would give me directions. I could hardly wait for the evening.

Around 4:00 PM, my cousin told me to get ready, anticipating heavy traffic. Just as I was prepared to leave, he got called to attend a major function and asked me to use his vehicle to drive myself to the Chinese Embassy.

I arrived at the Embassy at around 5:50 PM. At the gate, my name was on the list. Security guards opened the gate and directed me to the Ambassador's Residence. Another guard took the car to the parking lot. The Ambassador welcomed me warmly. As we prepared to enter, Dr.

Dahn arrived, followed closely by Minister Gwenigale. It was amazing to see the SSS (Special Security Service), now known as the EPS (Executive Protection Service), taking tactical positions. My background in Homeland Security made me suspect the President would attend this special dinner.

The Ambassador, Minister Gwenigale, Dr. Dahn, and I were escorted to a room. A few minutes later, the Ambassador and Minister Gwenigale stepped out. Dr. Dahn whispered that the President had arrived. Soon, the sound of sirens filled the air, signaling her arrival along with her sister, Aunt Jenni, and a few others. We were then led into a grand dining room where a beautifully set round table awaited us, complete with name tags for the seating arrangement. I sat to the left of the Ambassador, with the President on his right, followed by Minister Gwenigale. Dr. Dahn sat to my left, with others arranged around the table.

We enjoyed a nine-course meal while discussing various China-Aid projects in detail. President Sirleaf expressed her deep appreciation to the Chinese Ambassador and the Government of China for fulfilling the promises made by the President of China, particularly in Education (Fendell), Agriculture (CARI), and Health (Tappita Hospital). At the end of the dinner, the President made a toast to the Government and people of China and then to my arrival. The Ambassador followed suit.

As the President and others departed, the Ambassador asked me to stay a little longer. Once everyone had left, he took me to his private room, offering additional drinks and sweets. He began bombarding me with questions, starting with, "Dr. Kateh, what would you do if a new hospital were to be turned over to you?"

Realizing I was essentially in an interview, I composed myself and asked about the hospital's bed capacity and the status of its equipment. He handed me a list and confirmed the equipment was installed and ready. I discussed the human resources needed, the caliber of staff required, and how to capitalize on the investment to benefit not only the people of Liberia but also those in the region, including Ivory Coast and Guinea. After almost two hours of discussion, he thanked me and said he looked forward to working with me. His words puzzled me since I had just been hired for a new job in Tompkins County, New York.

Driving back to my cousin's house, I couldn't stop pondering the conversation. I arrived home at 11:00 PM. My cousin was up, waiting for me. "How did it go?" he asked.

I replied, "The meeting with the Chinese Ambassador? It was okay. He just wanted my professional opinion on how to effectively open the new hospital built by them."

He laughed. "The President didn't say anything to you about the hospital?"

"No," I responded.

"Wait and see," he said. "So, what's next?"

"Dr. Dahn told me we would leave for Tappita, Nimba County, at 10:00 AM tomorrow, so I need to be at the Ministry of Health and Social Welfare by 9:30 AM."

"Okay, I'll get you there on time," he assured me. I thanked him and went to bed, contemplating what the next day would bring.

CHAPTER 37

Journey to Tappita, Nimba County

The next morning, my cousin drove me to the Ministry of Health, where I met Dr. Dahn, who arrived about an hour later. We departed for Tappita, which is about an hour from Ganta, where I had previously worked. Dr. Dahn's driver, the late Tamba (may his soul rest in perpetual peace), did an excellent job navigating the road, especially the part between Gbarnga and Ganta, riddled with what I called "tub holes" due to their size. We arrived in Tappita around 7 PM and immediately went to the Ambassador's guesthouse. Dr. Gwenigale (deceased) told me to choose my room since I was the special guest. I insisted he and Dr. Dahn take their rooms first, and I would be comfortable with whatever was left. He laughed and agreed.

While they were inspecting the rooms, I ran over to the magnificent structure built by the Chinese. I felt like a child in a candy store, full of excitement, moving quickly from room to room, inspecting the entire hospital. Upon returning to the guesthouse, I pondered how this magnificent building would be run and eventually went to bed, still reflecting on the long journey and the road conditions from Gbarnga to Ganta.

Jackson F. Doe Memorial Regional Referral Hospital on the day it was turned over to the government of Liberia

The next morning, Friday, July 24, 2010, the stage was set, and the people of Nimba, particularly those from Tappita District and Upper Grand Gedeh, were jubilant. It was a vibrant and colorful event. As the night passed and the day of the occasion arrived, it became clear why I had been invited by the President. I took it in stride and followed the unfolding activities.

Before President Ellen Johnson Sirleaf's arrival to commence the turnover protocol for the multimillion-dollar facility, the seating arrangement was made. On the immediate left of the President was the Chinese Ambassador, H.E. Zhou Yuxiao, and on her immediate right was

Dr. Walter Gwenigale, the Minister of Health, followed by me. This arrangement almost confirmed my hypothesis. The Ambassador congratulated me for my willingness to run the hospital, which shocked me, but I thanked him.

After all the speeches, the contractor turned over the symbolic key to the Ambassador, who then passed it to the President, who then handed it to the Minister of Health and finally to me. It was an exciting experience. Luckily, my prior visit to the hospital came in handy. The President congratulated me and asked me to take her on a tour of the facility. Since I had already seen the equipment, it was easier for me to explain the various units and their functionalities.

At the end of the tour, the President, Ambassador, Minister of Health, and I went to the President's temporary residence. There, the President reiterated her appreciation for my accepting the challenge to run the hospital. The Ambassador asked if I had my passport because he wanted me to travel to Beijing, China, the following week for a leadership program for hospital administrators from twenty-seven African countries.

As we walked to the Ambassador's residence, he expressed his displeasure about the hospital's renaming. The original plaque read "China-Liberia Aid Hospital," but President Sirleaf had renamed it Jackson F. Doe Memorial Regional Referral Hospital. The Ambassador was curious about who Jackson F. Doe was and why the hospital was named after him.

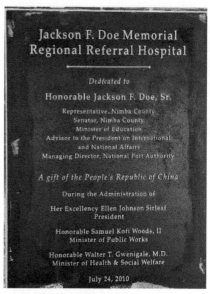

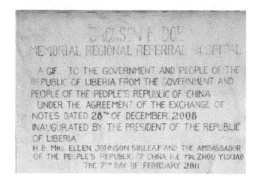

The various plaques at Jackson F. Doe Memorial Regional Referral Hospital

I explained the historical significance of Jackson Fiah Doe, a son of Nimba who had played a pivotal role in Liberia as Minister of Education and a Senator. I emphasized his contributions as the Standard Bearer of the Liberian Action Party and the subsequent events leading to the Civil War. I also highlighted the conciliatory gesture between the Gios of Nimba County and the Krahns of Grand Gedeh, noting that the late President Samuel K. Doe had initially appealed to the Government of Taiwan to build a hospital in Tappita to serve both regions.

Arial view of Jackson F. Doe Memorial Regional Referral Hospital three years later 2014

My explanation seemed to satisfy the Ambassador, who thanked me and entered his apartment, asking his Special Assistant to get my passport. The Assistant handed me his business card and instructed me to call him on the following Tuesday, July 27, 2010, for the departure date and ticket details.

By 9:00 AM on Saturday, July 24, 2010, the President and the convoy of dignitaries departed Tappita for Ganta. It was a slow process as the President stopped at every town where children had gathered. Dr. Gwenigale suggested we leave the convoy and find a place in Ganta to spend the night. Participating in the Tappita Hospital program was a moving experience.

Given the rapidly unfolding events and my acceptance of the position offered by the President, I wrote the following letter to Commissioner Anita Fitzpatrick, explaining why I could not assume the position in Tompkins County, New York:

Commissioner Anita Fitzpatrick,

It is with immense sadness that I write this letter. Since the beginning of my professional life, spanning over 14 years, I have never been faced with such an intense ambivalence regarding the making of a decision about where I would prefer to work.

During my interview, I informed the Search Committee that when I accepted the position in Anson County, I told them that I would have stayed a minimum of three years. In adherence to my words, I have spent over six years with the Department; hence, I was looking at a very long-term commitment with Tompkins County, New York.

I am still very excited about the opportunity to work in Tompkins County, NY. I am internally grateful to the Search Committee and the Tompkins County Government for the tremendous overreaching efforts to ensure that I become a member of the Tompkins County Government in order to contribute to making a difference in the lives of the residents of the County.

I received a call from the President of the Republic of Liberia, Her Excellency Madam Ellen Johnson Sirleaf, requesting that I kindly serve as the Medical Director (MD) and Chief Executive Officer (CEO) of a newly constructed 200-bed Referral Hospital in Tappita, Nimba County, Republic of Liberia.

As I compose this letter, I am filled with emotions of gratitude for the efforts you made to assimilate me with the Tompkins County Government, but due to the massive brain drain in my Country due to the ended civil crisis, Republic of Liberia, I have decided to accept the position in Liberia in order to help the country move forward.

Again, let me express my sincere thanks and appreciation for the efforts made on my behalf by the County. Most importantly, I wish you success in your endeavors to identify another suitable candidate for the position of Public Health Director. If there is any assistance that I can contribute to the establishment of the non-for-profit that I enumerated during my interview, please feel free to contact me.

Sincerely yours,

Francis N. Kateh, MD, MHA, MPS/HSL
Public Health Director, Anson County, North Carolina

Reflecting on my journey and the unfolding events, I knew that answering this call was the right decision for both my career and the people of Liberia.

CHAPTER 38

———·◆◆◆◆◆·———

Returning to Liberia

Upon returning to Wadesboro, Anson County, North Carolina, I found myself having restless nights, eagerly anticipating my return to Liberia. After attending to all administrative and personal matters in my office, I boarded a flight back to Liberia. Upon arrival, I immediately reported to the Ministry of Health, where I met with Dr. Bernice Dahn, the Chief Medical Officer of Liberia, followed by the late Dr. Walter Gwenigale, the Minister of Health. Informing them of my arrival to take on the challenging assignment at the Jackson F. Doe Memorial Regional Referral Hospital, Dr. Gwenigale, affectionately known as "Dr. G," welcomed me warmly and expressed his gratitude for my return. He encouraged me to provide "Free Care Services" to the people in the area. I quickly informed him that running such a facility on a "Free Care Basis" was impossible due to the necessity of sustainability and the inadequacy of government allotments. This discussion was somewhat contentious, but I had great respect for Dr. G, whom I called Daddy, so I acquiesced while planning to implement a cost recovery scheme.

After meeting with Dr. G, it was crucial to return to the hospital and assess what was in place to develop an operational plan. However, there was no logistical support, making it difficult to go to Tappita as planned.

The next three months were a nightmare. We couldn't get anything done, and I was under the impression that the Chinese Government would bring in a team to work with me to get the hospital operational.

During this period, I lived with a generous friend and his family, Dr. and Mrs. Adams Lincoln, who provided for me in every way. I will always remain grateful for their kindness. I spent over three months at their residence before moving to my place. Once I moved, it became more challenging to continuously go to the Ministry and engage the various stakeholders to plan for the hospital's start.

Realizing I wasn't making progress, I requested an audience with His Excellency the Ambassador of China, Zhou Yuxiao. He granted me an appointment and was excited to see me. My main agenda was to inquire about the Chinese Government's plan for opening the hospital and whether a team would be coming to work with me. The Ambassador asked if I had read the Memorandum of Understanding (MOU). When I said no, he called his secretary to make a copy for me.

The MOU revealed that there was no provision for staff to help operate the facility. It clearly stated an eighteen-month timeline from groundbreaking to completion, listing equipment and supplies necessary for smooth operation. I was shocked and realized that the assumptions about the Chinese Government's role were incorrect. I thanked the Ambassador for sharing the MOU and promised to get back to him with a plan to operationalize the hospital, ensuring it wouldn't become a "white elephant."

That night, I woke up at 2 AM and began brainstorming various options to ensure the hospital would provide quality services. I knew I needed to create a Mission Statement, Vision, Values, and Motto for the hospital. I drafted several statements to guide the hospital's direction.

Successful organizations thrive on their mission, vision, and values, which form the compass guiding their direction. I presented these statements at the first board meeting chaired by Dr. Gwenigale, who approved them and suggested I assist other health institutions in developing their mission statements. I also began working on a Personnel Policy to guide staff conduct and a Customer Service Training course mandatory for all staff. These documents were completed, setting a strong foundation for the hospital.

Setting a date for the official opening was stressful due to political pressure from the impending 2011 Election. The opposition labeled the hospital a white elephant, the community wanted jobs, and the County's interest was both economic and healthcare provision. Many vested interests pressured the hospital's opening. It was the first major hospital opening managed by a Liberian, as the John F. Kennedy Memorial Hospital had substantial USAID support for five years. The responsibility weighed heavily on me.

Recognizing that the hospital needed trained personnel and specialists to operate the equipment to avoid sending a negative message, I approached Prof. Dr. Nii Otu Nartey, the CEO of Korle-Bu Teaching Hospital in Accra, Ghana. He welcomed me and introduced me to Prof. Dr. Afua Adwo Jectey Hesse, the Chief of Medical Staff, who agreed to help in Obstetrics/Gynecology, Internal Medicine, Surgery, Radiology, and Laboratory. Encouraged, I returned to Liberia, eager to operationalize the hospital.

Back in Liberia, I faced challenges with the Ministry's bureaucracy. Cynical remarks about the hospital's readiness were common. Despite drafting an MOU between Korle-Bu and the hospital, the Ministry's legal office delayed its approval. Out of frustration, I sent the draft to Korle-Bu's CEO, who quickly approved it. I shared this with Hon. Vivian Cherue, Deputy Minister for Administration, who was thrilled. Meanwhile, without a vehicle, I was still expected to be in Tappita.

Just as we were wrapping up the MOU, I got a call from President Ellen Johnson Sirleaf, inquiring about the hospital's status. I briefed her on the progress but mentioned the significant challenges we faced: funding, logistics, and staff recruitment. She was visibly frustrated by the delay and informed me that she would call a meeting with the Minister and me the next day.

Immediately after my conversation with the President, I contacted Dr. Gwenigale to inform him about the upcoming meeting. He was furious, insisting that only he or the Chief Medical Officer had the authority to discuss the hospital with the President. He refused to attend the meeting, which left me in a precarious position.

The following morning, I made my way to the Ministry of Foreign Affairs, waiting for the President's call. By 10:00 AM, I was summoned to her office. She called Dr. Gwenigale, and with Mr. Patrick Sendolo present, the conversation began. True to his word, Dr. Gwenigale refused to attend, asserting that discussions should happen at the Ministry, not in the President's office.

The President instructed Mr. Sendolo to work with me to ensure the hospital opened by the third week of February. She didn't explain why this specific timeframe was crucial, but it was clear that this deadline was non-negotiable.

The next day, Dr. Gwenigale organized a meeting at the Ministry of Health. Mr. Sendolo, Deputy Minister Hon Vivian Cherue, and consultant Bill Martin were all present. Armed with a road map and a timeline, I was ready to meet the President's deadline. However, when we discussed the procurement of supplies necessary for the hospital's operation, Bill Martin insisted that the process would take at least three months. Frustrated by his condescending attitude, I finally let my pent-up anger spill out. Everyone was shocked; no one had ever challenged Bill Martin like that. He was a trusted figure at the Ministry, having a significant influence on major decisions.

Interestingly, this wasn't my first encounter with Bill Martin. He was a former manager of a large HMO in South Carolina that went bankrupt, leading to tragic consequences for several physicians. He had come to Liberia as a Christian missionary, working closely with Dr. Gwenigale at Phebe Lutheran Hospital. Their mutual trust was solidified over the years.

Mr. Bill Martin had quite a history. He lived in South Carolina and managed one of the largest HMOs in the state. Unfortunately, the entity went bankrupt, and tragically, a few physicians even committed suicide as a result. This led him to leave the United States in the nineties and come to Liberia as a Christian missionary. He worked with Dr. Gwenigale at Phebe Lutheran Hospital, which helped build a strong mutual trust between them. Eventually, he returned to the United States and settled in South Carolina. However, when the election was held in 2005 and Dr. Gwenigale became Minister of Health and Social Welfare, he called on his old friend Bill to come back as a Senior Advisor to the office of the Minister.

My first major encounter with Mr. Martin was in 2008 during a meeting at the office of the World Bank in Washington, DC. They were trying to source funding to establish a decentralized health system. I attended the meeting as an observer when Mr. Tornalah Varplah, then Deputy Minister for Planning and Research, made a presentation. During the Q&A session, a delegate from Finland asked about the process of controls, accountability, and reporting if the decentralized system was approved and funded. The response wasn't impressive, so I decided to explain how a decentralized health system functioned, drawing from my experience as a Public Health Director in North Carolina. After my brief explanation, the Finnish representative asked, "Why are you not in Liberia to run this system?" I replied simply, "I will in the future."

Dr. Gwenigale quickly added, "Dr. Kateh is a young Liberian that we are all proud of and know that he will come home one day to contribute to the building of the health system."

Those words resonated with me. After leaving the meeting, I reached out to Dr. Gwenigale and asked if there was something I could do to help. He responded positively and wanted me to work with Mr. Martin. He gave me Bill's contact information, and within two days, I received a call from him.

"Hi Francis, if I may call you Francis? This is Bill. Please call me Bill."

I responded, "Mr. Martin."

He insisted, "No, I said call me Bill."

So, I said, "What's up, Bill?"

He laughed and said, "I got a call from Dr. Gwenigale who wanted me to reach out to you to see if you could work on a decentralization plan and what would be the process for effective implementation and accountability." He added, "Francis, I have read so much about you in terms of delivering tangible results in the health sector. The work you did at Ganta Hospital as a young man is very incredible, and now what you are doing in North Carolina as a County Public Health Director is noteworthy."

I thanked him for the kind words but asked if there was already a plan for implementation and accountability. He said, "Yes, Mr. Varplah is working on something, but I haven't seen it yet. If you could put something together and share it with me, I will look at it and give you a call."

I told him, "Fine, but give me two weeks to put something together."

He responded, "That is excellent. I was thinking a month."

After our discussion, I returned to North Carolina and began drafting a decentralization plan, taking into consideration the Donabedian model of an effective health care system: Structure, Process, and Outcome. The plan needed to address several questions. What would be the communication chain for reporting from the community level up to the Central Level? Who are the stakeholders that could be held accountable if there is a breach in the system? What processes would be put in place to ensure an appreciable, tangible outcome?

After working on the concept, I sent the document to Bill and copied Dr. Gwenigale. A day or two later, Bill called me and wanted to meet in person to discuss the draft document. I drove to South Carolina with my family to meet him. Upon arrival, Bill gave us a tour of his condominium, which was very nicely located right on the beach in a small community that seemed quite exclusive. I noticed there were no black residents or customers at the restaurant where we had our discussion. After making some changes to the zero-draft document, Bill indicated he would share the document with Minister Varplah and, after getting inputs, would send it back for consensus on its cost and implementation. He also expressed interest in having me serve as an extended evaluator and consultant. I appreciated his thoroughness and then left.

However, after that meeting, Bill went silent. I sent numerous emails over the months that followed but received no response. This prompted me to become very aggressive with Bill when he later adopted a condescending attitude towards my President and my country during the planning stages of the Jackson F. Doe Memorial Hospital. When he walked out of the meeting, he later apologized, and I admired him for that. That moment seemed to clear the air and eliminate the dark cloud over the hospital's opening.

By the end of the week, Dr. Bernice Dahn called to inform me that a vehicle had been secured for the hospital—a Toyota Hilux, double cabin. With mobility secured, I was more determined than ever to ensure the hospital would be operational on schedule.

CHAPTER 39

————·◆◆◆◆◆·————

Recruitment of Staff for the Jackson F. Doe Memorial Regional Referral Hospital

Before my appointment, the Ministry had already recruited some staff for the hospital. This crucial detail had somehow been omitted during the entire process of preparing the hospital for operation. I had assumed I was in total control of managing the hospital, but this revelation was a significant wake-up call.

Realizing the hospital was nearly operational, I approached the Director of Human Resources at the Ministry to initiate the process of recruiting key clinical and support staff. To my surprise, he informed me that the recruitment had already been completed. I inquired about the process and learned that recommendations came from various authorities within the host county and the central office, appointing specific individuals to positions—a clear case of patrimony. This was another hurdle I had to overcome. The challenge was to get the hospital functional without alienating the host county and the central office. I needed to find a balance.

I requested the list of the already hired staff and discovered that some were already drawing salaries, even though the hospital hadn't opened yet. The list included 167 staff members, comprising clinical staff like nurses, aides, midwives, laboratory technicians, and administrative staff, including an administrator and deputy administrator.

Armed with this information, I engaged key individuals, such as the Chief Nursing and Midwife Officer, who had nominated several staff members. I asked if she knew these health workers personally and could vouch for their professional capabilities. She admitted that many names were given to her, and she had not personally spoken with them. I suggested we invite them for interviews and asked her to send a representative to assist with the process. She agreed, which was a significant hurdle to overcome. We formed a small committee and began the recruitment process. One positive outcome was identifying a competent Nursing Director who played a pivotal role in ensuring we recruited staff ready to begin work on day one.

With the recruitment underway, I needed to know the available funds for personnel, goods, and services. Dr. Dahn informed me of a $1.5 million budget based on projections for Tellewayon Hospital, a model used by the Swiss who had renovated and operated it for some time in Voinjama, Lofa County. Armed with this information, I developed an operational budget covering staff, medication, fuel, and lubricants for the three 750KVA generators, food for patients, cleaning materials, curtains, and vehicles (operational and ambulance). The budget was housed at the Ministry, which was another hurdle.

Realizing the need for specialized technicians and physicians, I finalized an MOU with Korle-Bu Teaching Hospital in Ghana. They sent a team for two months to build the capacity of our staff, particularly in diagnostics (radiology and laboratory). The team included a surgeon, an obstetrician/gynecologist, an internist, a radiology technician with expertise in CT scans, and a laboratory technician. They took sabbaticals from their institution and were paid based on a negotiated salary. With their support, we set the opening date for February 7, 2011, which was postponed to February 12, 2011.

Housing for staff was another critical issue. Several visits ensured everything was ready. I frequently traveled between Monrovia and Tappita to oversee preparations. On the day we transported staff to Tappita, we encountered an accident near Todee junction. Our staff spontaneously triaged the wounded and used ambulances to transfer them to C. H. Rennie Hospital in nearby Kakata. This incident assured me of the team's readiness.

We faced another challenge with living quarters for staff. We turned one of the inpatient facilities into a dormitory as there were no housing units built for nurses and support staff. Despite the Chinese Government's significant investment, only three duplexes were built—two one-bed, two two-bed, and two three-bedroom residences for doctors. We temporarily housed staff in part of the ward, which later became the Obstetrics/Gynecology Ward, until we found permanent accommodations.

Renovating the old, war-damaged Tappita Government Hospital provided a solution. The Chinese contractors had done minor repairs on a small portion of the facility to host staff during construction. We quickly renovated and painted a few rooms for staff to move in. As we hired additional staff, we continued renovations until the entire hospital was restored, using approximately 335 bundles of zinc. Portions of the old hospital were converted into storage for medication and general supplies.

As CEO/Medical Director, I continually conducted SWOT analyses to ensure the right steps were taken and investments yielded optimal dividends. Analyzing situations, prioritizing, and reprioritizing were crucial for success. It was an amazing and inspirational experience, grounded in the belief that relying on God's guidance was key. Our team of prayer warriors, led by the Nursing Director, continually sought divine guidance for the benefit of the people.

The contract with Korle-Bu Teaching Hospital was finalized, and we prepared for the official opening. Continuous engagement with the Chinese Ambassador, who became a friend, yielded dividends. I requested training for staff in China and established a tripartite relationship for consultants and technical support. These requests were approved and announced at the hospital's opening.

As we began providing care, the hospital's reputation spread rapidly. Patients traveled from Monrovia despite the bad road from Gbarnga to Tappita. However, the success came with its own set of challenges.

The official opening date was set for February 7, 2011, but was postponed due to the President's schedule. Interestingly, on February 6, 2011, the granddaughter of the Paramount Chief, who had given the land for the hospital, was brought in after 48 hours of obstructed labor. She gave birth to a baby girl named after President Ellen Johnson Sirleaf. This marked the beginning of patient admissions, leading to the official opening on February 12, 2011.

CHAPTER 40

———·◆◆◆◆◆·———

Official Opening and Dedication of the China-Liberia Aided Hospital (Jackson F. Doe Memorial Regional Referral Hospital) and Its Management

L-R, Dr. Kateh, H.E. President Sirleaf, H. E. Yuxiao, and Foreign Minister Mclintosh
Official Dedication of the Jackson F. Doe Memorial Regional Referral Hospital

The official opening and dedication of the Jackson F. Doe Memorial Regional Referral Hospital was an event of great significance, marking a milestone in Liberia's healthcare development. On February 12, 2011, during the ceremony, Ambassador Zhou Yuxiao spoke on behalf of China, reaffirming their commitment to aiding Liberia in its reconstruction efforts. He congratulated the hospital staff for transforming the building into a functional hospital with their skills and dedication. Ambassador Zhou also announced that China would train 25 Liberian medical personnel to effectively use and maintain the hospital's modern equipment as part of a tripartite agreement involving China, Egypt, and Liberia, a proposal I had put forward.

H. E Zhou Yuxiao, Ambassador People's Republic of China accredited to Liberia

When President Ellen Johnson Sirleaf cut the ribbon, the inscription on the marble stone read: "The Jackson F. Doe Memorial Referral Hospital, a gift to the People of the Republic of Liberia from the Government and People of the People's Republic of China under the agreement of the exchange of notes dated 28 December 2008 inaugurated by the President of the Republic of Liberia, H.E. Ellen Johnson Sirleaf and the Ambassador of the People's Republic of China, H.E. Zhou Yuxiao, the 7th day of February 2011."

The dedication ceremony was attended by prominent Liberians, including Foreign Minister Toga G. MacIntosh, Internal Affairs Minister Harrison Kanwea, Senator Adolphus Dolo, Representative Edwin Gaye, and Liberia's Chief Medical Officer, Dr. Bernice Dahn. President Sirleaf paid homage to the late Jackson Fiah Doe, Sr., acknowledging his sacrifices for the country's development and democratic process. She recalled his courage, strength, commitment, and dedication to national duty, virtues that inspired the Liberian people to rise again, particularly after the 2005 election. She offered words of solace to his widow, children, and family, assuring them that his legacy lived on.

The hospital was set to provide a range of services, including obstetrics and gynecology, pediatrics, surgery, internal medicine, ophthalmology, ENT, and emergency care. Its state-of the- art diagnostic department featured CT Scan, an X-Ray Machine, Ultrasound, Echocardiogram, Electrocardiograph (ECG), Electroencephalograph (EEG), Fluoroscopy, and a C-Arm.

Managing the hospital presented its own set of challenges. The Ministry of Health controlled the funds, complicating the planning and maintenance of the facility. Initially, Dr. Benedict Kolee and I were the only two physicians covering multiple departments between us. The bureaucratic hurdles were immense, with every check needing the Comptroller's signature, requiring me to make frequent, arduous trips from Tappita to Monrovia. This process was both frustrating and inefficient, but I persevered, driven by my commitment to the hospital's success.

One day, a lady who worked with the Comptroller finally intervened, highlighting the unfairness of making me wait for hours. This marked a turning point, although I continued to face challenges. Eventually, I had the opportunity to discuss these issues directly with President Sirleaf during the hospital's first anniversary. I explained the inefficiencies and risks associated with the current system, emphasizing the need for financial autonomy to ensure efficient, transparent, and accountable operations. President Sirleaf responded positively, instructing the Finance and Public Works Ministers to create a budget line for the hospital and provide funds for pavement of the major highway to mitigate the dust problem.

The hospital's budget was increased from $1.5 million to $2.5 million, granting us the financial autonomy needed. At the end of the fiscal year, we managed to save over $300,000, which we reinvested in staff housing and other critical infrastructure improvements. We renovated destroyed residences and converted the old Tappita Government Hospital into staff quarters. We also constructed a six-bedroom legislative residence and continued to enhance the hospital's facilities.

Part of the Destroyed Residence from the War

Demolished and redesigned residence for Doctors and other staff

The Old Tappita Government Hospital

Renovated Old Tappita Government Hospital and converted to staff quarters

H. E Ellen Johnson Sirleaf, Dr Louis Sullivan and Dr Kateh, during Dr Sullivan visit to Liberia for the ground breaking for a Medical School in Tappita, Nimba County

The Legislative Building

In addition to demolishing and constructing residences for doctors and other critical staff, we also built a six-bedroom house with four bathrooms, a large living and dining room, and a spacious kitchen. This guest residence, named "Legislative Residence," was made possible thanks to the approval of funds by the members of the 53rd Legislative body. Although it is unfortunate that the master plan for the hospital wasn't fully realized as envisaged, I remain grateful to God for what He has accomplished through me and the dedicated team working with me.

We focused on landscaping the hospital grounds and constructing eight residences in duplex form. These included one-bedroom self-contained units, two-bedroom units with two bathrooms, three-bedroom units, and five-bedroom units. Additionally, we designed an annex with fifteen private rooms and one presidential suite complete with an office, kitchen, security room, and other amenities. Unfortunately, the cost for this construction was never approved, and the master plan for transforming the hospital into a full medical community was left incomplete.

We had also procured ten acres of land for the construction of a medical school, a project for which the former Secretary of Health and Human Services, of the United States of America, Dr. Louis Sullivan, broke ground. Regrettably, this project did not come to fruition due to "politricks," a topic I plan to define and explore in detail in my next book.

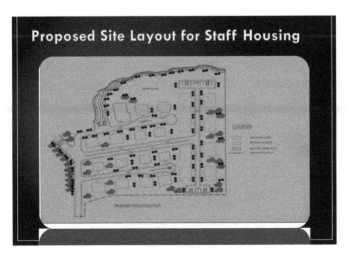

Proposed Site Layout for Staff Housing

Total View of the Schematic drawing of the Annex

"Caring with Compassion and Quality"   17

Proposed Private Annex with a Presidential Suite

In other words, the operation of the hospital was not a trial-and-error endeavor. Our plan and strategy were carefully discussed and implemented with the full support of the team. This was not a one-man show; it required the collective efforts of everyone, from the hygienists to the physicians and all the support staff, to elevate the hospital to the highest standard in the provision of care. Together, we brought to life the hospital's motto: "Caring with Compassion and Quality."

CHAPTER 41

———— ·◆◆◆◆◆·· ————

Response to the Semi-Autonomous Status

Following my one-on-one discussion with President Ellen Johnson Sirleaf and the justification I provided for granting the Jackson F. Doe Memorial Regional Referral Hospital semi-autonomous status, I knew I had to honor the trust she placed in me. The mission and vision of the hospital became my guiding principles:

Mission: To provide healthcare services to all through compassionate care and excellence in quality, to promote health education and research, and to strive to maintain national and international standards.

Vision: To be the hospital of choice in Liberia, recognized for having the most satisfied patients, the best possible clinical quality and outcomes, and the best physicians and employees.

With these goals in mind, I faced the enormous task of recruiting the right experts with the necessary credentials. I began by scouting around the sub-region, but the quotes I received for hiring specialists were beyond our allotted budget. Determined to find a solution, I contacted a friend in Uganda for assistance, but the applications I received were from young specialists who still needed guidance. Refusing to give up, I sent information to Ethiopia, where I received over forty applications across various needed disciplines.

After reviewing and scrutinizing the applications, I set off for Addis Ababa, Ethiopia. I made the necessary contacts and booked a hotel to serve as an office for conducting intensive interviews. Over the course of several days, I interviewed twenty-three specialists and selected the six we needed: a general surgeon, a pediatrician, an obstetrician/gynecologist, a radiologist, an internist, and a pathologist. As I write this autobiography, two of those original six are still working tirelessly to bring relief to the people of Liberia and its environs, fulfilling my pledge to the President.

The Minister of Public Works kept the President's promise and paved two kilometers of road in front of the hospital, significantly reducing the dust entering our facility. This improvement was crucial in maintaining a clean and healthy environment for our patients and staff.

To meet the hospital's primary objective of providing quality and patient-centered health services, we embarked on landscaping and developing a park to aid in the healing process of our patients. We continued constructing additional housing units to support the growing staff as the hospital reached full capacity. As the hospital's reputation grew, it became imperative to expand our services and build international connections with other specialists and organizations. Collaborations with entities like the Korle-Bu Neuro Foundation in Canada and the National Medical Association (the oldest Black medical association with over 40,000 members) led to significant milestones, such as performing the first Craniotomy, Laminectomy and Transurethral Resection of the Prostate (TURP) in Liberia.

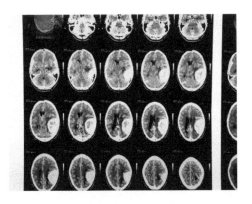

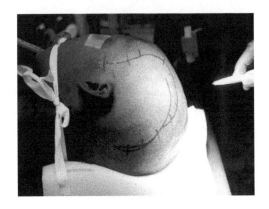

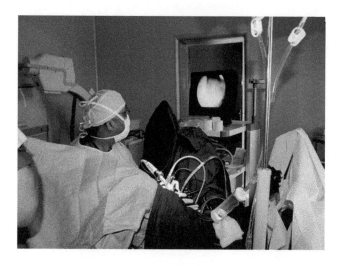

With the recruitment of the appropriate staff, the hospital became unstoppable in delivering quality care with a patient-centered focus. We averaged over twenty patients from Monrovia daily despite the poor road conditions from Gbarnga to Tappita. Patients also came from Liberia's southeastern region and neighboring countries like Guinea and Côte d'Ivoire. The hospital became a household name in Liberia and its environs, known for its commitment to excellence and compassionate care.

The journey to operationalize and elevate the Jackson F. Doe Memorial Regional Referral Hospital was challenging but incredibly rewarding. With every hurdle, we grew stronger, always keeping the mission and vision at the forefront of our efforts. The success of the hospital was a testament to the dedication and hard work of every team member, from the hygienists to the physicians. Together, we created a beacon of hope and quality healthcare in Liberia, embodying our motto: "Caring with Compassion and Quality."

CHAPTER 42

------- ·◆◆ ◆ ◆◆· -------

What Made Jackson F. Doe Memorial Regional Referral Hospital Different

Jackson F. Doe Memorial Regional Referral Hospital was a generous gift from the People's Republic of China to Liberia. Given the past failures to revitalize the John F. Kennedy Memorial Center, I knew a different leadership approach was necessary. Drawing on my educational background and practical experience, I decided to operate the hospital using a corporate model. Hence, I requested from the President of Liberia to be both the Chief Executive Officer and Medical Director. The criteria set for this role included being a medical doctor with a master's in health administration or a doctor with a master's in business administration and at least five years of experience in a proactive healthcare environment. This setup allowed for a holistic view of the administration, emphasizing strategic business modules for sustainability while focusing on the cardinal function of a hospital—patient care.

My combined background as a physician and a healthcare administrator gave me a unique perspective on running an effective and efficient healthcare system. These traits, blended with servant leadership, became the cornerstone of the hospital's success. *Servant leadership,* as defined by Webster's Dictionary, is a leadership philosophy where the leader's goal is to serve. Unlike traditional leadership, where the leader's focus is on the thriving of the company or organization, servant leadership focuses on the team's growth and well-being. The theory posits that when team members feel personally and professionally fulfilled, they produce high-quality work more efficiently and productively. This management style, where the leader serves the employees, is particularly suitable for non-profit organizations.

As a trained and experienced health administrator, I sought to find a niche in my leadership style that suited the environment and vision of the hospital. Changing the mindset of staff, especially regarding patient care, was a significant challenge. I emphasized at every monthly meeting that everyone at the hospital, from hygienists to doctors, was equally important and deserved respect. I often said, "We are all on this train called Jackson F. Doe Memorial Regional Referral Hospital, and everyone on this train has a specific role and responsibility. If anyone neglects their role, the train will derail."

Despite these encouragements, some hygienists, previously known as cleaners before the Ebola crisis, felt ashamed of their jobs despite earning decent wages by Liberian standards. Their reluctance to take pride in their work led to repeated warnings, and eventually, I had to dismiss five of them for failing to keep the hospital clean. This action met resistance from key community members, but I stood firm, refusing to be swayed by bribes or gifts. I hoped new hires would change their work attitude, but the same trend emerged.

I decided to lead by example, participating in cleaning and maintaining the hospital to ensure a hospitable environment free of nosocomial infections. Landscaping was also a challenge, as the topsoil had been removed to level the land for construction. I planted grass seeds and flowers to beautify the area.

One Saturday morning, a patient and her husband arrived after a long journey from Monrovia, delayed by their vehicle getting stuck. When they reached the hospital, they found me planting flowers. The husband didn't believe it was me until I greeted him by name. They were both shocked but impressed. I examined the lady, diagnosed her, and scheduled surgery for Monday. After the successful surgery, she fully recovered.

The young landscaper I trained was initially teased by other workers, but I instilled in him the importance of landscaping and the beauty it adds. As time passed, he began to appreciate his work, receiving praise as the flowers took shape. I also enlisted the Liberia Forestry Authority to plant indigenous trees, enhancing the hospital's grounds. Today, the landscaping at Tappita stands as a testament to this innovative approach.

To make the environment more conducive for patients and visitors, we developed a park managed by a refugee who carved various animals to beautify the space. The park became a popular spot for families and even for couples taking wedding photos.

When the Ebola crisis hit, Dr. Walter Gwenigale asked me to provide my expertise in Margibi County. My efforts led to significant achievements, resulting in my elevation to Deputy Incident Manager for Medical, Planning, and Research. Working with a dedicated team and partners, we performed miracles, gaining community trust and saving countless lives. After Liberia was declared Ebola-free on May 9, 2015, I was appointed Chief Medical Officer/Deputy Minister for Health Services by President Ellen Johnson Sirleaf. This role was beyond my dreams, but it became a reality through dedication and divine guidance.

CHAPTER 43

What Was the Harvest?

Returning home was no easy task; it tested every fiber of my existence. However, one thing I will always cherish is the realization that God made me for a purpose. Despite not being born in a hospital, health center, or clinic, my mother's situation became the impetus for my journey into healthcare, transforming adversity into a mission to deliver healthcare at various levels. This journey took a toll on my family and resulted in significant financial loss. Yet, what gives me hope is my choice to pursue public service grounded in personal values rather than market values. It attracts those who want to "make a difference" for others and seek to maximize their self-worth rather than their net worth. This was a personal quote I shared during my acceptance of an Honorary Doctorate in Public Service from MacMurray College, Jacksonville, on May 16, 2016.

Dr. Francis Kateh '91
DOCTOR OF PUBLIC SERVICE

◇

MacMurray College salutes Francis N. Kateh, M.D., physician, administrator, MacMurray alumnus, and courageous leader of his homeland's successful fight against the Ebola virus.

Dr. Kateh's dedication to healing humanity began early, when he volunteered at Ganta United Methodist Hospital in his native land of Liberia. After attending the University of Liberia, he came to MacMurray with a scholarship sponsored by the Central Illinois Conference of the United Methodist Church, earning a bachelor's degree in biology here in 1991.

He obtained his medical degree at Spartan University School of Medicine in St. Lucia in the Caribbean. And then, when Liberia was involved in its first civil war, Dr. Kateh returned to his homeland to help alleviate the widespread suffering at great personal risk to his life.

Back in the United States, Dr. Kateh prepared for further service in times of crisis by earning a certificate in Healthcare Leadership and Administrative Decision Making from the Federal Emergency Management Agency's Center for Domestic Preparedness; a Master of Professional Studies in Homeland Security Leadership at the University of Connecticut in Mansfield, Connecticut, and a Master of Health Administration from Governors State University in University Park, Illinois.

He returned to Liberia to help rebuild the healthcare system after the civil wars. And then, as Deputy Incident Manager for Medical Response, Kateh found himself at the forefront of the fight against the Ebola virus that infected more than 10,000 Liberians and killed 4,000. He volunteered to take a trial Ebola vaccine, again at risk to his own safety, hoping it would encourage others in the epidemic zone to follow his example. The U.S. Centers for Disease Control recognized the strategies Dr. Kateh developed to fight Ebola as being the most effective and efficient.

MacMurray has already recognized Dr. Kateh's exceptional merit with our Distinguished Alumni Award for the Millennium in 2000 and our Distinguished Young Alumni Award in 2001. Now that he has accomplished so much more, he deserves our highest honor.

Turning down the position of Public Health Director in Tompkins County, Ithaca, New York, to accept an unknown role in Liberia was a leap of faith. Despite the uncertainty, I was convinced that the power above all powers would see me through. This decision, though it led to a threatened and eventual divorce, remains one I do not regret. The transformation of lives and the establishment of competitive healthcare standards in postwar Liberia were worth the sacrifice. The protection and ingenuity granted to me through God's grace are immeasurable. I am humbled and grateful to be of service to the people of Liberia serving two Presidents (H. E. Ellen Johnson Sirleaf and H.E. George M. Weah) as the Chief Medical Officer/Deputy Minister of Health Services (6/2015 -1/2024), someone who was never born in a health facility... isn't that the grace of God, hence to Him alone be the glory and honor.

REPUBLIC OF LIBERIA

THE PRESIDENT

EJS/MOS/RL/343/2015   May 22, 2015

Hon. Dr. Francis N. Kateh
Deputy Minister/Chief Medical Officer
Ministry of Health
Republic of Liberia

Dear Hon. Kateh:

I am pleased to advise that based upon notification from the Honorable Liberian Senate of your confirmation, you are hereby appointed **Deputy Minister/Chief Medical Officer**, Republic of Liberia, effective Tuesday, May 21, 2015.

It is expected that you will take office immediately and if not already submitted, you are required to file, by June 9, 2015, with the Anti-Corruption Commission the Declaration of Income, Assets and Liabilities.

Please accept my congratulations and my expression of trust in your ability to make a meaningful contribution in your area of responsibility as we strive to move our country forward in a process aimed at enhancing peace, reconciliation and development.

Sincerely,

Ellen Johnson Sirleaf

REPUBLIC OF LIBERIA

THE PRESIDENT

GMW/MOS/RL/269/2018   February 13, 2018

Hon. Dr. Francis Nah Kateh
c/o Ministry of Health
Monrovia, Liberia

Dear Hon. Kateh:

I am pleased to inform you of your appointment as **Deputy Minister for Health Services and Chief Medical Officer, Ministry of Health,** effective February 6, 2018, Republic of Liberia.

Your appointment is based on your desire to play a more meaningful role in the promotion of peace, reconciliation and development of our country and in recognition of your loyalty and commitment to your country.

I trust that you will carry out your responsibilities with utmost diligence and dedication.

Kind regards.

Sincerely,

George M. Weah

"CHANGE FOR HOPE"

An email from Mr. Stephen Estes, former Deputy Human Resources Director of Tompkins County, years after I turned down the appoint-

ment, epitomizes my love for my country and my commitment to following God's direction and my late mother's guidance:

Good afternoon, Dr. Kateh:

I am not sure this email address is good any longer, but I was thinking about you, and the challenges you must be facing in Liberia these days. Therefore, I thought I would just touch base and say hi. I was at a conference the other day and for some reason your name came up, which made me start wondering how things are going for you. The last I knew; you were going to work for the Chinese to assist them in running the Liberian medical system. I often wonder how that panned out for you. I also worry about the Ebola situation and hope that you, your family, and all of those under your direction are as safe as can be. Anyway, I don't really know why I write other than to let you know that, even though you were a fleeting moment in my life, you made your mark on my memory and remain in my thoughts. I will always respect that you made a commitment to your home country when you could have just as easily taken the position with us. Peace be with you, Dr. Kateh. I hope all is well.

Sincerely, Stephen

The most significant harvest is making a difference in the lives of others. God has taken me from humble beginnings to be a blessing to others, embodying the principle of "not for self but others." For this, I remain blessed.

SHALOM!

References

Mbiti John S. 2006. African Religions & Philosophy 2Nd rev. and enl. ed. repr ed. Oxford: Heinemann.

Ozioma, Ezekwesili-Ofili, Nwamaka Chinwe, Okaka. "Herbal Medicines in African Traditional Medicine" In Herbal Medicine, edited by Philip Builders. London: IntechOpen, 2019. 10.5772/intechopen.80348.

Soderlund Walter C. 2008. Humanitarian Crises and Intervention: Reassessing the Impact of Mass Media. Sterling VA: Kumarian Press

Milton Keynes UK
Ingram Content Group UK Ltd.
UKHW040256291024
450401UK00006B/77

9 781962 467391